Practical and Efficient
SAS® Programming
The Insider's Guide

Martha Messineo

sas.com/books

Contents

About This Book

What Does This Book Cover?

This book does not teach a novice how to program with SAS. Its purpose is to provide SAS programmers with better tools for accomplishing a variety of tasks using DATA steps, macros, the SQL procedure, and other procedures.

It covers data manipulation, tricks for writing better programs, utilities to simplify application development, and a variety of techniques, including using recursion, using lookup tables, and reading metadata.

Is This Book for You?

The expected audience for this book is SAS programmers with some—or a lot of—experience in writing SAS code. You should have some experience with the DATA step (basic concepts and the use of functions) and with the macro facility (macro variables and simple macro programming). In addition, a basic understanding of PROC SQL and other standard procedures is useful.

What Should You Know about the Examples?

This book includes examples that you can run for yourself to gain hands-on experience with SAS. All of the examples are self-contained. Most of the data that is used is from the standard sashelp sample tables (for example, sashelp.class and sashelp.cars). A few examples create their own data tables.

Software Used to Develop the Book's Content

Base SAS 9.4 was used to develop the content for this book. Some of the code works in any version of SAS, some only works in SAS 9, and some only works in SAS 9.4. In addition, there is one example that requires access to the SAS Metadata Repository, formerly known as the *SAS Open Metadata Repository*, so the SAS Metadata Server must be installed to use these examples.

The code was primarily developed on a Windows 7 platform and runs on any operating system that SAS runs on. However there are some examples that are written specifically for Windows and/or Linux and would need to be modified to run on a different operating system.

Example Code and Data

You can access the example code for this book by linking to its author page at https://support.sas.com/authors.

All the data used in the example programs is available in the standard sashelp library available to all SAS users or is created by the example program.

We Want to Hear from You

SAS Press books are written *by* SAS Users *for* SAS Users. We welcome your participation in their development and your feedback on SAS Press books that you are using. Please visit sas.com/books to do the following:

- sign up to review a book
- recommend a topic
- request information on how to become a SAS Press author
- provide feedback on a book

Do you have questions about a SAS Press book that you are reading? Contact the author through saspress@sas.com or https://support.sas.com/author_feedback.

SAS has many resources to help you find answers and expand your knowledge. If you need additional help, see our list of resources: sas.com/books.

About The Author

Martha Messineo is a principal software developer in the SAS Solutions OnDemand division at SAS. She has more than 30 years of SAS experience, most of which she has spent writing SAS programs in a variety of positions, including those in Technical Support, Professional Services, Management Information Systems, SAS Solutions OnDemand, and Research and Development. In her Research and Development role, she worked with SAS Data Integration Studio, IT Resource Management, and SAS Asset Performance Analytics. Before coming to SAS, she worked for MetLife Insurance in the Capacity Planning and Performance Tuning division, where she used SAS to analyze mainframe performance data. She is a SAS Certified Base Programmer for SAS9, a SAS Certified Advanced Programmer for SAS9, and a Data Integration developer for SAS9.

Learn more about this author by visiting her author page at http://support.sas.com/Messineo. There you can download free book excerpts, access example code and data, read the latest reviews, get updates, and more.

Acknowledgments

I want to thank some people for all their SAS wisdom!

- Thanks to Robert Woodruff, whose questions inspired me to write this book—if Robert has a question about something, then someone else probably has the same question.
- Thanks also to Scott Schwartz for all his questions and his side projects.
- Thanks to my team at MetLife for introducing me to SAS: Sue Rosansky, John Grasing, Wing Louie, Fred Duca, John Brobst, Chris Fallotico, Tom Grimason, Chris Szuter, Bruce Geddes, and Richard Ondrovic.
- Thanks to my mentor at SAS, Brian Perkinson, for teaching me so much and, more importantly, for being a great friend.
- Thanks to my fellow SAS programmers for all the tricks that I have "borrowed" from them over the years: Sally Walczak, Kevin Hobbs, Randy Poindexter, Jim Ward, Dave Berger, Kathy Wisniewski, Jason Sharpe, Merry Rabb, Greg Smith, Carrie Santiago, Jim Hart, Chuck Bass, Chris Watson, Chris Lysholm, Stephanie Watkins, Carl Sommer, and Tim Rowles. I have been so lucky to have worked with so many outstanding SAS programmers.
- Thanks to my teachers (many years ago) at RPCS who taught me that I could be a computer nerd and relatively articulate at the same time.
- And many thanks and love to my husband, Chris Messineo, for encouraging me to take on this project and then putting up with all the time I spent on it. And thanks to our two "kids," Buddy and Lily, who didn't get as many walks as they deserved because I was busy writing.

Chapter 1: My Favorite Functions

Introduction

There are literally hundreds of DATA step functions. It is so easy to get overwhelmed by the sheer volume that you miss can some real gems. Here are some functions that I can't live without and some different ways to use them.

Concatenating Strings

If you need to combine two or more strings into a single string, you can use the concatenation operator: ||. For example, you can take two strings such as "Mary" and "Smith" and create a new string with a value of "Mary Smith":

```
newString = "Mary" || " " || "Smith";
```

Tip: You might see SAS code that uses !! instead of || as the concatenation operator. Both are valid operators. The reason that some people use the !! is because several years ago (okay … many years ago), mainframe keyboards didn't have the | character. So, if you see !! instead of ||, it probably means that the coder originally learned SAS on a mainframe.

The CAT() functions—CAT(), CATS(), CATX(), CATQ(), and CATT()—can also be used to concatenate strings together. So why not just use the standard ‖ operator to do concatenation? The CAT() functions do additional work including stripping leading and trailing spaces and converting numbers to characters. I still use the ‖ syntax at times, but I use the CAT() functions more.

So if you are doing basic concatenation, you can simplify the following syntax:

```
newString = trim(left(string1)) || left(string2);
```

Instead, you can use the following syntax, which is much easier to type and read:

```
newString = cats(string1, string2);
```

CATX() adds delimiters between your strings. You can simplify the following syntax:

```
newString = trim(left(string1)) || " " || left(string2);
```

Instead, you can use this:

```
newString = catx(" ", string1, string2);
```

Converting Numbers to Characters

The ability of these functions to do basic concatenation makes them great tools, but you can do more.

The CAT() functions also convert numeric values into character strings, so you don't have to use a PUT() function. So to concatenate a number to a string, you could do this:

```
newString = trim(left(string1)) || left(put(number, 10.));
```

But this is much simpler:

```
newString = cats(string1, number);
```

When converting a number to a string without using the CAT() function, you need to use the PUT() function and specify the format that you want. So you might do this:

```
newString = left(put(number, 3.));
```

But what happens if you miscalculate, and the value for number is bigger than 999? SAS software fits the value into 3 characters, so the result is "1E3" instead of "1000". Using the CATS() function, you don't have to worry about the length of the actual number, and you can just do the following:

```
newString = cats(num);
```

The example in Program 1.1 and SAS Log 1.1 shows how to convert a number using the traditional style, and then using the CATS() function.

Program 1.1: Converting a Number to a String

```
data _null_;
   length newString_put newString_cat $10;
   number = 1000;

   newString_put = left(put(number, 3.));
   putlog newString_put=;

   newString_cat = cats(number);
   putlog newString_cat=;
run;
```

SAS Log 1.1: Converting a Number to a String

```
newString_put=1E3
newString_cat=1000
```

Tip: If you are converting a number to a character and you need the new string to look a certain way—maybe with a $ or commas—you need to use a PUT() function with the appropriate format. Use the CAT() functions only to convert simple numbers into simple strings.

Adding Delimiters

At some point, you might have had to create a string that consists of a series of other strings separated by a delimiter (such as a blank or a comma). When you do this without CATX(), you need to use syntax similar to this:

```
data _null_;
   set sashelp.class end = eof;
   where sex eq "F";

   length girlList $500;
   retain girlList "";

   if (_n_ eq 1) then
      girlList = left(name);
   else
      girlList = trim(left(girlList)) || ", " || left(name);

   if (eof) then
      putlog girlList=;
run;
```

This syntax produces the following output:

```
girlList=Alice, Barbara, Carol, Jane, Janet, Joyce, Judy, Louise,
Mary
```

This code does what it should: It puts all the girls' names in a comma delimited list. But it needs additional code to avoid putting an extra comma at the beginning (as in this example) or end of the list.

An alternative is to use CATX(). Then you don't have to worry about extra commas, because CATX() only puts the delimiter between non-blank values. Program 1.2 and SAS Log 1.2 show how to use the CATX() function to add a comma between the girls' names.

Program 1.2: Concatenating Strings with a Delimiter

```
data _null_;
   set sashelp.class end = eof;
   where sex eq "F";

   length girlList $500;
   retain girlList "";

   girlList = catx(", ", girlList, name);

   if (eof) then
      putlog girlList=;
run;
```

SAS Log 1.2: Concatenating Strings with a Delimiter

```
girlList=Alice, Barbara, Carol, Jane, Janet, Joyce, Judy, Louise,
Mary
```

Another way that CATX() is useful is when you want to concatenate several strings with a delimiter, but you don't want extra delimiters if some of the strings are blank. For example, suppose you have a first, middle, and last name, and you want to create a full name out of the parts. If the middle name is blank, you don't want to add extra blanks. If you use ||, then you must check to see whether the middle name is blank, or you will end up with too many blanks in the name.

The code in Program 1.3 creates a data table to use with the example in Program 1.4, and Figure 1.1 shows a listing of the table.

Program 1.3: Creating the Names Data Table

```
data names;
   input @1 first $5. @7 middle $5. @15 last $10.;
   datalines;
Sue           Jones
Ann    K.     Smith
Joe           Thomas
Sally Jo      Anderson
;
```

Figure 1.1: Creating the Names Data Table

Obs	first	middle	last
1	Sue		Jones
2	Ann	K.	Smith
3	Joe		Thomas
4	Sally	Jo	Anderson

You could just concatenate the 3 names like this:

```
name = strip(first) || " " || strip(middle) || " " || strip(last);
```

You would get these names as a result:

```
Sue  Jones
Ann  K.  Smith
Joe  Thomas
Sally  Jo  Anderson
```

You can see that both Sue and Joe have an extra space because they don't have middle names.

To deal with this issue, you can test the middle name to see whether it is blank before adding it to the name:

```
if (middle eq "") then
   name = strip(first) || " " || strip(last);
else
   name = strip(first) || " " || strip(middle) || " " ||
          strip(last);
```

The names then have the correct spacing:

```
Sue Jones
Ann K. Smith
Joe Thomas
Sally Jo Anderson
```

Tip: You could use the COMPBL() function to take care of the extra spaces. It replaces multiple spaces with a single space.

Using CATX(), you don't have to worry about extra blanks. If a value is missing, CATX() doesn't add the extra delimiter, as you can see in the example in Program 1.4 and SAS Log 1.3.

Program 1.4: Using CATX() to Avoid Checking for a Blank Value

```
data _null_;
   set names;
   length name $50;
   name = catx(" ",
               first, middle, last);
   putlog name=;
run;
```

SAS Log 1.3: Using CATX() to Avoid Checking for a Blank Value

```
name=Sue Jones
name=Ann K. Smith
name=Joe Thomas
name=Sally Jo Anderson
```

Using the OF Shortcut

In addition to all the benefits of the CAT() functions, you can also use the OF shortcut to concatenate a series of strings.

If you want to create a unique key variable on each record of the table, then you can concatenate all the variables with a delimiter between the values. Or maybe you want to create a comma-delimited file out of all the variables in the table. If you don't use CATX(), then it is a complicated task, as the following program illustrates:

```
proc contents data = sashelp.class
               out = contents (keep = name varnum)
               noprint nodetails;
run;

data _null_;
   set sashelp.class;
   length newString $1000;
   newString = "";
   do p = 1 to nobs;
      set contents point = p nobs = nobs;
      newString = strip(newString) || "," ||
                   strip(vvaluex(c_name));
   end; /* do p - loop through contents */
   /* get rid of beginning comma */
   newString = substr(newString, 2);
run;
```

Tip: If you have the name of a variable in another variable, use the VVALUEX() function to get the first variable's value:

x = "y";
y = "abc";
z = vvaluex(x);
putlog x= y= z=;

Output: x=y y=abc z=abc

Program 1.5 and SAS Log 1.4 show how to use the CATX() function with the OF _ALL_ shortcut to simplify adding all the variable values together.

Program 1.5: Using CATX() to Concatenate All Strings

```
data _null_;
   set sashelp.class;
   length newString $1000;
   newString = catx(",", of _ALL_);
   putlog newString=;
run;
```

SAS Log 1.4: Using CATX() to Concatenate All Strings

```
newString=Alfred,M,14,69,112.5
newString=Alice,F,13,56.5,84
newString=Barbara,F,13,65.3,98
newString=Carol,F,14,62.8,102.5
newString=Henry,M,14,63.5,102.5
newString=James,M,12,57.3,83
newString=Jane,F,12,59.8,84.5
newString=Janet,F,15,62.5,112.5
newString=Jeffrey,M,13,62.5,84
newString=John,M,12,59,99.5
newString=Joyce,F,11,51.3,50.5
newString=Judy,F,14,64.3,90
newString=Louise,F,12,56.3,77
newString=Mary,F,15,66.5,112
newString=Philip,M,16,72,150
newString=Robert,M,12,64.8,128
newString=Ronald,M,15,67,133
newString=Thomas,M,11,57.5,85
newString=William,M,15,66.5,112
```

Removing Leading and Trailing Spaces

When you need to remove spaces from the beginning and end of a string, the STRIP() function can take care of it for you. It is a great function because it replaces two functions: LEFT() and TRIM(). How could you not love that?

The STRIP(*string*) function is almost the same as TRIM(LEFT(*string*)). The only difference is that if you use STRIP() on a blank string, it returns a value of zero characters, while TRIM() returns a string of one blank character. Using TRIMN(LEFT(*string*)) returns a zero-length string just as the STRIP() function does.

Program 1.6 and SAS Log 1.5 compare the results from using the LEFT() and TRIM() functions to using the STRIP() function.

Program 1.6: Removing Spaces

```
data _null_;
   x1 = " abc   ";
   x2 = "      ";
   array x {2};
   do i = 1 to dim(x);
      putlog x{i}=;
      y = "*" || x{i} || "*";
      putlog "with spaces " y=;
      y = "*" || trim(left(x{i})) || "*";
      putlog "trim(left()) " y=;
      y = "*" || trimn(left(x{i})) || "*";
      putlog "trimn(left()) " y=;
      y = "*" || strip(x{i}) || "*";
      putlog "strip " y=;
      putlog;
   end;
run;
```

SAS Log 1.5: Removing Spaces

```
x1=abc
with spaces y=* abc   *
trim(left()) y=*abc*
trimn(left()) y=*abc*
strip y=*abc*

x2=
with spaces y=*     *
trim(left()) y=* *
trimn(left()) y=**
strip y=**
```

The STRIP() function is great, and you should be using it.

Finding Non-Blank Values

There might be times when you have a set of variables that are either blank or not, and you want to find the one that is not blank. For example, you might have a first-name variable and a nickname variable, and you want to use the nickname unless it is blank. Otherwise, you will use the first name. To do this, you could use IF-THEN-ELSE. However, there is a function that can do it for you: COALESCE(). It is very useful in both PROC SQL and the DATA step.

In PROC SQL it is invaluable when joining tables. If the same column is in multiple tables, you can use the COALESCE() function to get the one that has a non-blank value. In addition, you can set a default if all the values are blank.

The code in Program 1.7 creates two sample data tables that are used in the Program 1.8 example, and Figure 1.2 and Figure 1.3 are listings of those tables.

Program 1.7: Creating Two Tables

```
data table1;
   input x y;
   datalines;
1 1
3 .
5 5
6 6
;
data table2;
   input x y;
   datalines;
1 10
2 20
4 40
7 .
;
```

Figure 1.2: Creating Two Tables—table1

Obs	x	y
1	1	1
2	3	.
3	5	5
4	6	6

Figure 1.3: Creating Two Tables—table2

Obs	x	y
1	1	10
2	2	20
3	4	40
4	7	.

You can join them with PROC SQL and use a CASE statement to get the *x* and *y* values set correctly. That works, but is a lot of work (especially if you can't always remember the CASE statement syntax):

```
proc sql;
   create table newTable as
      select t1.x as t1_x, t2.x as t2_x,
             t1.y as t1_y, t2.y as t2_y,
             case
                when (t1.x ne .) then t1.x
                when (t2.x ne .) then t2.x
                else 0
             end as x,
             case
                when (t1.y ne .) then t1.y
                when (t2.y ne .) then t2.y
                else 0
             end as y
         from table1 as t1
              full join
            table2 as t2
               on t1.x eq t2.x;
   quit;
```

Program 1.8 shows how you can use the COALESCE() function without bothering with the CASE statement, and Figure 1.4 is a listing of the joined data.

Program 1.8: Using COALESCE() in an SQL Join

```
proc sql;
   create table newTable as
      select t1.x as t1_x, t2.x as t2_x,
             t1.y as t1_y, t2.y as t2_y,
             coalesce(t1.x, t2.x, 0) as x,
             coalesce(t1.y, t2.y, 0) as y
         from table1 as t1
              full join
              table2 as t2
                 on t1.x eq t2.x;
quit;
```

Figure 1.4: Using COALESCE() in an SQL Join

Obs	t1_x	t2_x	t1_y	t2_y	x	y
1	1	1	1	10	1	1
2	.	2	.	20	2	20
3	3	.	.	.	3	0
4	.	4	.	40	4	40
5	5	.	5	.	5	5
6	6	.	6	.	6	6
7	.	7	.	.	7	0

The COALESCE() function also works in a DATA step and helps you get rid of some IF statements. Program 1.9 creates a table to use in the example in Program 1.10, and Figure 1.5 is a listing of the table created.

Program 1.9: Creating a Table with Missing Character Values

```
data table;
   input @1 a $1. @3 b $1.;
   datalines;
A B

  C
D
;
```

Figure 1.5: Creating a Table with Missing Character Values

Obs	a	b
1	A	B
2		
3		C
4	D	

Suppose you want to create a new column that is either the value from a or the value from b, whichever is not blank. If they are both blank, then set the value to "Z". You could use an IF-THEN-ELSE statement like this:

```
if (a ne "") then
    newVar = a;
else if (b ne "") then
    newVar = b;
else
    newVar = "Z";
```

Or you can use the COALESCE() function and avoid all that typing, as shown in Program 1.10 and SAS Log 1.6.

Program 1.10: Handling Blank Values with COALESCE()

```
data _null_;
    set table;
    newVar = coalescec(a, b, "Z");
    putlog a= b= newVar=;
run;
```

SAS Log 1.6: Handling Blank Values with COALESCE()

```
a=A b=B newVar=A
a= b= newVar=Z
a= b=C newVar=C
a=D b= newVar=D
```

Tip: If you are coalescing character values in a DATA step, you must use the COALESCEC() function; the COALESCE() function is only for numeric values. However, you can use COALESCE() in PROC SQL for either numeric or character values.

And finally, you can use the COALESCE() function to set a missing value to a constant in a single statement. So the following code can be replaced:

```
if (string eq "") then
    string = "BLANK";
```

Instead, you can do this:

```
string = coalescec(string, "BLANK");
```

I have a friend who likes the coalesce() function so much that he asks prospective SAS programmers to explain it in interviews.

Creating Datetime Values

If you have a SAS date value (stored as the number of days since January 1, 1960) and a SAS time value (the number of seconds since midnight), and you need to create a SAS datetime value (the number of seconds since January 1, 1960), you can use the DHMS() function.

The DHMS() function takes a SAS date value, an hour value (the hour part of a time), a minute value (the minute part of a time), and a second value (the seconds part of a time), and returns a SAS datetime value. Here is the syntax:

```
datetime = dhms(date, hour, minute, second);
```

Program 1.11 and SAS Log 1.7 show that if you only have a date value, DHMS() can also be used to make a SAS datetime value, using zeros for the hour, minute, and second arguments.

Program 1.11: Creating a Datetime Value with No Time Value

```
data x;
    date = "01Jan2016"d;
    datetime = dhms(date, 0, 0, 0);
    putlog date= date9.;
    putlog datetime= datetime16.;
run;
```

SAS Log 1.7: Creating a Datetime Value with No Time Value

```
date=01JAN2016
datetime=01JAN16:00:00:00
```

What about when you have a date and a time? I have often done this to create a datetime value:

```
datetime = dhms(date,
                hour(time),
                minute(time),
                second(time));
```

This code works, but there is some extra effort (both typing and processing) to split out the hour, minute, and second values from the time. However, because a SAS time is stored as the number of seconds since midnight, and because the fourth argument to DHMS() is seconds, you just need to use the time variable as the seconds argument with zeros for the hours and minutes, as you can see in Program 1.12 and SAS Log 1.8:

Program 1.12: Creating a Datetime Value with Date and Time Values

```
data _null_;
    date = "01Jan2016"d;
    time = "13:25"t;
    datetime = dhms(date, 0, 0, time);
    putlog date= date9.;
    putlog time= time8.;
    putlog datetime= datetime16.;
run;
```

SAS Log 1.8: Creating a Datetime Value with Date and Time Values

```
date=01JAN2016
time=13:25:00
datetime=01JAN16:13:25:00
```

Creating Macro Variables

If you work much with macros and macro variables, you have probably used CALL SYMPUT() to create a macro variable from a DATA step value. You might have even used CALL SYMPUTX() to create a local or global macro variable—to ensure that the macro variable is created with the correct macro scope. What you might not realize about CALL SYMPUTX() is that it can make a macro variable out of a numeric DATA step variable, so you don't have to convert the value to a character.

Without using CALL SYMPUTX(), you have probably done something like this:

```
call symput("macroVar", trim(left(put(numVar, 10.1))));
```

And, if you are familiar with the CAT() functions, you can simplify this:

```
call symput("macroVar", cats(numVar));
```

Program 1.13 and SAS Log 1.9 show that with CALL SYMPUTX(), you can simplify the code even more.

Program 1.13: Creating a Macro Variable with CALL SYMPUTX()

```
data _null_;
   numVar = 14.5;
   call symputx("macroVar", numVar);
run;
%put &=macroVar;
```

SAS Log 1.9: Creating a Macro Variable with CALL SYMPUTX()

```
MACROVAR=14.5
```

The CALL SYMPUTX() routine can also help you with character strings. It trims the leading and trailing blanks before assigning the string to the macro variable, so you don't need to use TRIM(LEFT()) or STRIP() to make sure that your macro value isn't padded with blanks.

Finding Words

The SCAN() function is a great tool for splitting a string into chunks using some sort of delimiter. For example, if you have a name such as "Mary Jones", you can use SCAN() to pull off the first word "Mary" and the second word "Jones" to store them in separate variables:

```
firstName = scan("Mary Jones", 1, " ");
lastName = scan("Mary Jones", 2, " ");
```

If you want to get the last word or words from a string, there are lots of ways to do it. You can use a loop and keep looping until you get to the end of the string:

```
count = 0;
lastWord = "";
do until(word eq "");
   count + 1;
   word = scan(longString, count, " ");
   if (word ne "") then
      lastWord = word;
   else
      leave;
end; /* do until - end of string */
```

You can reverse the string, take the first word, and then reverse that word:

```
word = reverse(scan(reverse(strip(longString)), 1, " "));
```

You can use the COUNTW() function to scan the last word:

```
word = scan(longString, countw(longString, " "), " ");
```

And I'm sure you can think of other ways to do this. However, the easiest way is to use the feature of the SCAN() function that allows you to scan from the end of a string. This feature was introduced in SAS 8. All you need to do is use a negative counter; this tells SCAN() to start at the end of the string. So use -1 to get the last word, -2 to get the second-to-last word, and so on. If you know the length of the string and the length of the substrings, then you can just use the SUBSTR() function to pull out the last word. But if you don't have that information, the SCAN() function can handle it for you. Program 1.14 and SAS Log 1.10 show how to get the last and the second-to-last values from a string.

Program 1.14: Finding the Last Word with a Negative Counter

```
data _null_;
   longString = "Word1 Word2 Word3 Word4";
   putlog longString=;

   length wordLast wordSecondLast $10;

   wordLast = scan(longString, -1, " ");
   putlog wordLast=;

   wordSecondLast = scan(longString, -2, " ");
   putlog wordSecondLast=;
run;
```

SAS Log 1.10: Finding the Last Word with a Negative Counter

```
longString=Word1 Word2 Word3 Word4
wordLast=Word4
wordSecondLast=Word3
```

The SCAN() function is incredibly useful, and has lots of parameters that you might have missed (as I did).

Counting Words

If you have a string that consists of values that are delimited with some character, you might want to know how many words are in that string, or you might want to do different processing based on the number of strings. For example, if you have a name string such as "Tom Jones" or "Sally Jo Smith", you need to check the number of words to determine whether there is a middle name or not before you separate the parts of the name.

The COUNTW() function can tell you how many words are in a string. A word is defined as a substring that is bounded by delimiters and the beginning and end of the string. COUNTW() saves you having to scan words from a string until you get a blank value, signaling that you have reached the end of the string, like this:

```
count = 0;
do until(word eq "");
   count + 1;
   word = scan(longString, count, " ");
   if (word ne "") then
      putlog count= word=;
end; /* do until - end of longString */
```

The example in Program 1.15 and SAS Log 1.11 shows how you can use the COUNTW() function to simplify your code.

Program 1.15: Using COUNTW() to Get the Number of Words

```
data _null_;
   longString = "Word1 Word2 Word3 Word4";
   length word $10;
   do count = 1 to countw(longString, " ");
      word = scan(longString, count, " ");
      putlog count= word=;
   end; /* do count - loop through words */
run;
```

SAS Log 1.11: Using COUNTW() to Get the Number of Words

```
count=1 word=Word1
count=2 word=Word2
count=3 word=Word3
count=4 word=Word4
```

The code in Program 1.16 and SAS Log 1.12 shows how to use COUNTW to split apart a name that might include a middle name and might not, you can use countw() to check the number of words to see whether there are 2 words (first and last names only) or 3 words (first, middle, and last names).

Program 1.16: Using COUNTW to Split a Name

```
data _null_;
   input name $20.;
   putlog name=;

   length firstName middleName lastName $20;
   firstName = scan(name, 1, " ");
   if (countw(name) eq 3) then
      middleName = scan(name, 2, " ");
   lastName = scan(name, countw(name), " ");

   putlog firstName= middleName= lastName=;
   datalines;
Mary Jones
Tom A. Smith
;
```

SAS Log 1.12: Using COUNTW to Split a Name

```
name=Mary Jones
firstName=Mary middleName= lastName=Jones
name=Tom A. Smith
firstName=Tom middleName=A. lastName=Smith
```

The COUNTW() function has plenty of arguments that enable you to specify the correct delimiters.

Replacing Substrings

The TRANSTRN() function and its sister (brother?), tranwrd() function, enable you to replace part of a string with another string. The primary difference between the two is that TRANSTRN() lets you remove part of a string without leaving a blank in its place.

> **Tip**: The TRANWRD() function was introduced in SAS 9.1, and the TRANSTRN() function followed in SAS 9.2. I got into the habit of using TRANWRD(), so I tend to use it most of time. If you haven't gotten into this habit, then ignore TRANWRD() and use TRANSTRN() instead; it has all the same functionality as TRANWRD(), plus the ability to remove a string without leaving a blank.

You can achieve the same results by using functions such as INDEX(), SUBSTR(), and TRANSLATE(), but TRANSTRN() is so much easier. For example, without using TRANSTRN() you could use INDEX() and SUBSTR() and a DO loop to replace all blanks with "???":

```
newString = string;
do until (i eq 0);
   i = index(strip(newString), " ");
   if (i gt 0) then
    newString = substr(newString, 1, i-1) || "???" ||
                substr(newString, i+1);
end;
```

And you can remove all instances of a string (in this case, "???") like this:

```
do until (i eq 0);
   i = index(strip(newString), "???");
   if (i gt 0) then
    newString = substr(newString, 1, i-1) ||
                substr(newString, i+3);
end;
```

Program 1.17 and SAS Log 1.13 show that you can use TRANSTRN() to simplify both of these tasks considerably.

Program 1.17: Replacing Parts of a String Using TRANSTRN()

```
data _null_;
   string = "This is a test";
   putlog string=;

   /* replace all blanks with "???" */
   newString = transtrn(string, " ", "???");
   putlog newString=;

   /* remove all "???" */
   newString = transtrn(newString, "???", trimn(""));
   putlog newString=;
run;
```

SAS Log 1.13: Replacing Parts of a String Using TRANSTRN()

```
string=This is a test
newString=This???is???a???test
newString=Thisisatest
```

Tip: When you want to remove part of a string without leaving a blank space in its place, use TRIMN("") instead of just "" as the replacement value in TRANSTRN(). This replaces the first string with a zero-length string.

Warning: If you want to use a variable instead of a quoted value for the second or third arguments in TRANSTRN() or TRANWRD(), you probably need to use STRIP() to remove leading and trailing blanks.

If you use an unstripped variable as the target (the second argument), the function looks for the value with all the padding on the end.

And if you use an unstripped variable as the replacement value (the third argument), the function adds all the blanks at the end of the value when it does the replace.

The TRANWRD() and TRANSTRN() functions are incredibly useful when you are cleaning and transforming data.

Using %SYSFUNC() to Run DATA Step Functions

If you haven't used the %SYSFUNC() macro function, you have been missing out on access to most of the SAS DATA step functions from outside the DATA step. With %SYSFUNC() you can run a DATA step function outside a DATA step.

Program 1.18 and SAS Log 1.14 show a simple example that uses %SYSFUNC() with the NOTDIGIT() function to check whether the value in a macro variable is a digit or not (0 indicates that it is a digit, and a number indicates the first position in the string of a non-digit).

Program 1.18: Using %SYSFUNC() with NOTDIGIT()

```
%let string = 1A2;
%let rc = %sysfunc(notDigit(&string));
%put &=rc;
%let string = 12;
%let rc = %sysfunc(notDigit(&string));
%put &=rc;
```

SAS Log 1.14: Using %SYSFUNC() with NOTDIGIT()

```
1    %let string = 1A2;
2    %let rc = %sysfunc(notDigit(&string));
3    %put &=rc;
RC=2
4    %let string = 12;
5    %let rc = %sysfunc(notDigit(&string));
6    %put &=rc;
RC=0
```

If you want to put a date value into a macro variable, you can save the numeric value of the date (which is the number of days since 01JAN1960) or you can save the formatted value as you can see in Program 1.19 and SAS Log 1.15. Note that you can apply a format to the value as the second parameter of %SYSFUNC().

Program 1.19: Using %SYSFUNC() with Dates

```
%let date = %sysfunc(today());
%put &=date;
%let date = %sysfunc(today(), date9.);
%put &=date;
```

SAS Log 1.15: Using %SYSFUNC() with Dates

```
1    %let date = %sysfunc(today());
2    %put &=date;
DATE=20877
3    %let date = %sysfunc(today(), date9.);
4    %put &=date;
DATE=27FEB2017
```

In Program 1.20 and SAS Log 1.16, you can see that %SYSFUNC() can be used to do lots of other work as well, including reading data tables.

Program 1.20: Using %SYSFUNC() to Read a Table

```
%macro readData;
   %let did = %sysfunc(open(sashelp.class (where=(sex='F'))));
   %if (&did eq 0) %then
   %do;
      %put %sysfunc(sysmsg());
      %return;
   %end;
   %do i = 1 %to %sysfunc(attrn(&did, NLOBSF));
      %let rc = %sysfunc(fetchobs(&did, &i));
      %let name = %sysfunc(getvarc(&did,
                         %sysfunc(varnum(&did, name))));
      %let age = %sysfunc(getvarn(&did,
                         %sysfunc(varnum(&did, age))));
      %put &=name &=age;
   %end;
   %let did = %sysfunc(close(&did));
%mend;
%readData;
```

SAS Log 1.16: Using %SYSFUNC() to Read a Table

```
NAME=Alice AGE=13
NAME=Barbara AGE=13
NAME=Carol AGE=14
NAME=Jane AGE=12
NAME=Janet AGE=15
NAME=Joyce AGE=11
NAME=Judy AGE=14
NAME=Louise AGE=12
NAME=Mary AGE=15
```

When using %SYSFUNC(), do not enclose the function parameters in quotation marks even if you would normally do that in a DATA step. For example, in a DATA step, you would use this syntax:

```
string = transtrn(string, "a", "A");
```

When using %SYSFUNC() outside a DATA step, the syntax would be as follows:

```
%let string = %sysfunc(transtrn(&string, a, A));
```

Another thing to be aware of is that you have to use a separate %SYSFUNC() for each function you call:

```
%let name = %sysfunc(getvarc(&did,
                    %sysfunc(varname(&did, 3))));
```

There are some functions that cannot be used with %SYSFUNC():

ALLCOMB()	LEXCOMB()
ALLPERM()	LEXCOMBI()
DIF()	LEXPERK()
DIM()	LEXPERM()
HBOUND()	MISSING()
IORCMSG()	PUT()
INPUT()	RESOLVE()
LAG()	SYMGET()
LBOUND()	variable information functions—VNAME(), VLABEL(), and the like

Instead of PUT() and INPUT(), you can use PUTN(), putc(), inputn(), and INPUTC().

Chapter 2: Data Tables

Introduction

Working with data tables is a fundamental part of SAS, which means there are many programming choices when you need to manipulate data tables. This chapter shows you some techniques that will help you work with data tables.

Copying Variable Attributes

I often find that I need to make a new table that has the same variables and attributes as an existing table, usually because I'm going to append the new table to the existing table and want the attributes to be the same. You can hardcode all the lengths, formats, informats, and labels for the variables, but this creates a maintenance headache if the original table changes. The technique I use is to add these lines at the beginning of the DATA step:

```
if (0) then
   set table;
```

Here is how this works:

- When you run a DATA step, the program data vector (PDV) is created at compile time, before any of the code in the DATA step is actually executed. The PDV is a list of all the variables and their attributes that are used during the DATA step. The compiler gathers all these variables from the DATA step code and from the headers of any tables that are referenced in the DATA step. The compile step happens one time before executing the DATA step.
- When you use this technique, the variables and their attributes in the table on the SET statement are brought into the PDV during the compile step.
- When the DATA step is executed after the compile step, and each time it runs the IF statement, it considers the condition to be false (because 0=false in SAS), so it doesn't execute the statement, or statements, after the THEN. This means that no data is ever read from that table.
- So the variables and their attributes are brought into the PDV by the compiler, but records in the table are never read.

In the example in Program 2.1, I want to create a new table with the same variables and attributes as the sashelp.class table.

Program 2.1: Creating a New Table with the Same Attributes

```
data newClass;
   if (0) then
      set sashelp.class;
   input name sex age height weight;
   datalines;
Susan F 10 50 75
Chris M 10 53 80
;
```

After running the code in Program 2.1, the variables in the newClass table are the same as the ones in the sashelp.class table, as you can see in Figure 2.1 and Figure 2.2.

Figure 2.1: sashelp.class Variable Properties

Figure 2.2: work.newClass Variable Properties

Be sure to put the IF statement close to the beginning of your DATA step; the variables are added to the PDV in the order that the compiler finds them.

Variable lengths can sometimes cause issues when you use this technique. If you use a LENGTH statement before the IF (0) statement, and you specify the length of a variable to be shorter than the variable in the table, then you will get warnings in your SAS log:

```
data newClass;
   length name $4;
   if (0) then
      set sashelp.class;
   input name sex age height weight;
   datalines;
Susan F 10 50 75
Chris M 10 53 80
;
```

This code causes these warnings in the SAS log:

```
WARNING: Multiple lengths were specified for the variable name by
input data set(s). This can cause truncation of data.
```

If your code is working the way you want, you can temporarily turn off these warnings in your logs with the VARLENCHK= option. You should then turn the warnings back on after the DATA step. I don't recommend setting the option at the beginning of your code, because there are times that the warning indicates an issue that should be fixed. But if you are sure that the warning is not a problem, you can set the option to NOWARN before the DATA step and then back to WARN after the DATA step as in the example in Program 2.2 and SAS Log 2.1.

Program 2.2: Suppressing Warning Messages

```
options varlenchk = NOWARN;
data newClass;
   length name $4;
   if (0) then
      set sashelp.class;
   input name sex age height weight;
   datalines;
Susan F 10 50 75
Chris M 10 53 80
;
options varlenchk = WARN;
```

SAS Log 2.1: Suppressing Warning Messages

```
1    options varlenchk = NOWARN;
2    data newClass;
3       length name $4;
4       if (0) then
5          set sashelp.class;
6       input name sex age height weight;
7       datalines;

NOTE: The data set WORK.NEWCLASS has 2 observations and 5 variables.
NOTE: DATA statement used (Total process time):
      real time              0.01 seconds
      cpu time               0.00 seconds


8    ;
9    options varlenchk = WARN;
```

Warning: When you use this technique without referencing any other input sources (with SET, MERGE, UPDATE or INFILE statements), then you need to use a STOP statement at the end of the DATA step. Even though the SET statement is not executed, the DATA step still knows that it is there, so the DATA step waits for an end-of-file marker to tell it to stop processing. And since no data is read from an input source, there will never be an end-of-file marker. So you must use the STOP statement to explicitly tell the DATA step to stop processing.

Reading Data with a SET Statement

Reading an existing data table into your DATA step is a simple process using the SET statement. You can also use the SET statement to read multiple tables and to read the data out of sequence.

Concatenating Tables

If you want to add multiple tables together, one after another, you can simply list all the tables on a single SET statement. This causes the DATA step to read all the records from the first table, then all the records from the second table, and so on until it has read all the records from all the tables. The tables don't even have to have the same variables, though usually it is more useful if they do.

The code in Program 2.3 creates two class tables—one for each year—that are used in the example in Program 2.4. Figure 2.3 and Figure 2.4 show these two tables.

Program 2.3: Creating class Tables

```
data class_2016;
   set sashelp.class (obs = 5);
run;

data class_2017;
   set sashelp.class (keep = name sex age
                      obs = 6);
   age + 1;
   grade = 100;
run;
```

Figure 2.3: Creating class Tables—work.class_2016

Obs	Name	Sex	Age	Height	Weight
1	Alfred	M	14	69.0	112.5
2	Alice	F	13	56.5	84.0
3	Barbara	F	13	65.3	98.0
4	Carol	F	14	62.8	102.5
5	Henry	M	14	63.5	102.5

Figure 2.4: Creating class Tables—work.class_2017

Obs	Name	Sex	Age	grade
1	Alfred	M	15	100
2	Alice	F	14	100
3	Barbara	F	14	100
4	Carol	F	15	100
5	Henry	M	15	100
6	James	M	13	100

You can use a SET statement to concatenate these two tables, even though they don't have the same variables. The example in Program 2.4 shows how to concatenate the two class tables, and Figure 2.5 shows a listing of the output table. I also use the IN= option to help create a year variable to indicate which table the record came from.

Program 2.4: Concatenating Tables with a SET Statement

```
data class;
   set class_2016 (in = in_2016)
       class_2017 (in = in_2017);
   if (in_2016) then
      year = 2016;
   else if (in_2017) then
      year = 2017;
run;
```

Figure 2.5: ConcatenatingTables with a SET Statement

Obs	Name	Sex	Age	Height	Weight	grade	year
1	Alfred	M	14	69.0	112.5	.	2016
2	Alice	F	13	56.5	84.0	.	2016
3	Barbara	F	13	65.3	98.0	.	2016
4	Carol	F	14	62.8	102.5	.	2016
5	Henry	M	14	63.5	102.5	.	2016
6	Alfred	M	15	.	.	100	2017
7	Alice	F	14	.	.	100	2017
8	Barbara	F	14	.	.	100	2017
9	Carol	F	15	.	.	100	2017
10	Henry	M	15	.	.	100	2017
11	James	M	13	.	.	100	2017

The Height and Weight variables are only in the class_2016 table, so the values are missing on the 2017 records. The grade variable is only in the class_2017 table, so the values are missing on the 2016 records.

Interleaving Tables

If you have multiple tables that are sorted by the same variables, then you can interleave the tables using a SET statement with a BY statement. The records are read in sorted order from all the input tables. Program 2.5 and SAS Log 2.2 show how to interleave a table containing the males in sashelp.class with the females from sashelp.class.

Program 2.5: Interleaving Tables

```
proc sort data = sashelp.class
           out = male;
   where sex = "M";
   by name;
run;

proc sort data = sashelp.class
           out = female;
   where sex = "F";
   by name;
run;

data _null_;
   set male
       female;
   by name;
   putlog name= sex=;
run;
```

SAS Log 2.2: Interleaving Tables

```
Name=Alfred Sex=M
Name=Alice Sex=F
Name=Barbara Sex=F
Name=Carol Sex=F
Name=Henry Sex=M
Name=James Sex=M
Name=Jane Sex=F
Name=Janet Sex=F
...
```

You can see that the records are read in sorted order from both tables, rather than all the records being read from the first table and then all the records from the second table.

Using Multiple SET Statements

You can use more than one SET statement in a single DATA step. If you do this, the DATA step reads a record from the data table each time the SET statement is executed. So if you have two SET statements one after another in the DATA step, it reads a record from the table on the first SET statement and then a record from the table on the second SET statement in the same iteration of the DATA step. This means that you must be careful about the variables in each table so that values from a second table don't overwrite those from the first table. The other thing to watch out for is the end of file. The DATA step stops iterating when it reaches the end of any file that is being read, so whichever table has the smallest number of records causes the DATA step to stop iterating before all the records in the other tables have been read.

Using NOBS= and POINT=

When I use multiple SET statements, I usually use the NOBS= and POINT= options on at least one of the SET statements.

- The NOBS= option creates a variable that contains the number of records in the incoming table.
- The POINT= option points to a variable that you create that contains the record number that is read from the table.

So if you are reading one table, and you want to look up values in a second table, you can use the POINT= and NOBS= options to let you loop through the second table. Using this technique means that you don't hit the end-of-file condition that stops your DATA step iteration. Even if you read the last record, it doesn't trigger the end-of-file flag.

Program 2.6 and SAS Log 2.3 are an example of using the POINT= and NOBS= options for looking up values in a second table. Using the sashelp.class table and 10 records from the sashelp.classfit table as a lookup table, you can look up the predicted weight for each person.

Program 2.6: Using POINT= and NOBS=

```
data classfit (keep = name predict);
   set sashelp.classfit (obs = 10);
run;

data _null_;
   set sashelp.class;
   do p = 1 to nobs;          ❶
      set classfit (rename = (name = name_fit))   ❷
         point = p nobs = nobs;
      if (name eq name_fit) then
      do;
         putlog name= weight= predict= 5.1;
         leave;       ❸
      end;
   end;
run;
```

❶ The p variable is used in the POINT= option and is incremented so that each record of the classfit table is read. The nobs variable is automatically set to the number of records in the classfit table by the NOBS= option.

❷ It is necessary to rename the name variable when reading the classfit table, because the name value from sashelp.classfit overwrites the name value from sashelp.class.

❸ The LEAVE statement causes the execution to exit the DO loop when it finds the match so that you don't read the entire lookup table for each record in the sashelp.class table. If you are looking for multiple matches, then you would not use the LEAVE statement.

SAS Log 2.3: Using POINT= and NOBS=

```
Name=Alice Weight=84 predict=77.3
Name=Carol Weight=102.5 predict=101.8
Name=James Weight=83 predict=80.4
Name=Jane Weight=84.5 predict=90.1
Name=Janet Weight=112.5 predict=100.7
Name=Jeffrey Weight=84 predict=100.7
Name=John Weight=99.5 predict=87.0
Name=Joyce Weight=50.5 predict=57.0
Name=Louise Weight=77 predict=76.5
Name=Thomas Weight=85 predict=81.2
```

I tend to use p as the variable name for the POINT= option, and nobs for the NOBS= option. These are easy to remember. If you are setting multiple tables using the POINT= and NOBS= options, then you should use a different variable for each POINT= and NOBS= option.

Using KEY=

The KEY= option enables you to read matching records from a second table using an index. The index must be set up on the second table and the variables included in the index must be set to appropriate values before executing the SET statement with the KEY= option.

Program 2.7 and SAS Log 2.4 use the previous example for POINT= and NOBS=, except this time I've added an index to the classfit lookup table so that I can use the KEY= option.

Program 2.7: Using KEY=

```
data classfit (index = (name)
               keep = name predict);
   set sashelp.classfit (obs = 10);
run;

data _null_;
   set sashelp.class;
   set classfit key = name;
   if (_iorc_ eq 0) then        ❶
      putlog name= weight= predict= 5.1;
   _error_ = 0;        ❷
run;
```

❶ The automatic variable, _IORC_, is set to a return code when you use a SET statement with a KEY= option. A value of 0 indicates that the record was found.

❷ If you are using a KEY= option, you should always set the automatic variable _ERROR_ to 0 at the end of the DATA step. Otherwise all the variable values are written to the SAS log whenever _IORC_ is not 0.

SAS Log 2.4: Using KEY=

```
Name=Alice Weight=84 predict=77.3
Name=Carol Weight=102.5 predict=101.8
Name=James Weight=83 predict=80.4
Name=Jane Weight=84.5 predict=90.1
Name=Janet Weight=112.5 predict=100.7
Name=Jeffrey Weight=84 predict=100.7
```

```
Name=John Weight=99.5 predict=87.0
Name=Joyce Weight=50.5 predict=57.0
Name=Louise Weight=77 predict=76.5
Name=Thomas Weight=85 predict=81.2
```

If you want to use the KEY= option to look up multiple records from the lookup table, you need to use the KEYRESET= option. This option indicates when to start at the top of the index and when to read the next value in the index. The KEYRESET= option points to a variable that you should set to 1 when you want to start at the top of the index, and set to 0 when you want to read the next value in the index. The option will very nicely set the variable to 0 each time it reads a record.

The example in Program 2.8 and SAS Log 2.5 looks up all the values associated with a given id value.

Program 2.8: Looking up Multiple Records with KEY=

```
data lookup (index = (id));
   input id $1. value $char10.;
   datalines;
1 Pear
1 Peach
1 Banana
2 Apple
3 Grape
3 Lemon
;
data main;
   input id $1.;
   datalines;
1
2
2
;
data _null_;
   set main;
   reset = 1;       ❶
   putlog id= reset=;
   do until (_iorc_ ne 0);
      set lookup key = id keyreset = reset;
      if (_iorc_ ne 0) then
         leave;
      putlog id= value= reset=;
   end;
   _error_ = 0;
run;
```

❶ You must set the reset variable to 1 to force the program to start at the top of the index. If you don't have reset=1, the program does not find any records when it tries to look up the second id=2 value, because it has already read the one record in the lookup table for id=2.

SAS Log 2.5: Looking up Multiple Records with KEY=

```
id=1 reset=1
id=1 value=Pear reset=0
id=1 value=Peach reset=0
id=1 value=Banana reset=0
id=2 reset=1
id=2 value=Apple reset=0
id=2 reset=1
id=2 value=Apple reset=0
```

Determining Which Table a Record Is From

If you use a SET statement or a MERGE statement with multiple data tables, you can take advantage of the IN= data set option. This option enables you to specify a variable that is automatically set to 1 (true) when the current record comes from that data table, and is set to 0 (false) when the current record does not come from that data table.

Program 2.9 and SAS Log 2.6 show an example of using the IN= option with an interleaving SET statement. Records are read from the tables in the order specified on the BY statement.

Program 2.9: Using IN= with SET Statement

```
proc sort data = sashelp.class
          out = male;
   where sex = "M";
   by name;
run;

proc sort data = sashelp.class
          out = female;
   where sex = "F";
   by name;
run;

data _null_;
   set male (in = inM)
       female (in = inF);
   by name;
   putlog inF= inM= name= sex=;
run;
```

SAS Log 2.6: Using IN= with SET Statement

```
inF=0 inM=1 Name=Alfred Sex=M
inF=1 inM=0 Name=Alice Sex=F
inF=1 inM=0 Name=Barbara Sex=F
inF=1 inM=0 Name=Carol Sex=F
inF=0 inM=1 Name=Henry Sex=M
inF=0 inM=1 Name=James Sex=M
inF=1 inM=0 Name=Jane Sex=F
inF=1 inM=0 Name=Janet Sex=F
inF=0 inM=1 Name=Jeffrey Sex=M
inF=0 inM=1 Name=John Sex=M
inF=1 inM=0 Name=Joyce Sex=F
```

```
inF=1 inM=0 Name=Judy Sex=F
inF=1 inM=0 Name=Louise Sex=F
inF=1 inM=0 Name=Mary Sex=F
inF=0 inM=1 Name=Philip Sex=M
inF=0 inM=1 Name=Robert Sex=M
inF=0 inM=1 Name=Ronald Sex=M
inF=0 inM=1 Name=Thomas Sex=M
inF=0 inM=1 Name=William Sex=M
```

The IN= option is particularly useful in the MERGE statement to determine whether the current record came from one or multiple tables. In fact, I rarely use a MERGE statement without also using the IN= option.

The example in Program 2.10 and SAS Log 2.7 shows merging two tables together. One table has all the students from the sashelp.class table who are 13 and older, and the second table has all the students who are 14 and older. This means that some of the records from the 13+ table do not have matches in the 14+ table. Using the IN= option on each table in the MERGE statement enables you to determine whether each record came from both tables or only one table.

Program 2.10: Using IN= with MERGE Statement

```
proc sort data = sashelp.class
          out = thirteen_plus;
   where age ge 13;
   by name;
run;

proc sort data = sashelp.class
          out = fourteen_plus (rename = (age = age14));
   where age gt 13;
   by name;
run;

data _null_;
   merge thirteen_plus (in = in13_plus)
         fourteen_plus (in = in14_plus);
   by name;
   putlog in13_plus= in14_plus= name= age=;
run;
```

SAS Log 2.7: Using IN= with MERGE Statement

```
in13_plus=1 in14_plus=1 Name=Alfred Age=14
in13_plus=1 in14_plus=0 Name=Alice Age=13
in13_plus=1 in14_plus=0 Name=Barbara Age=13
in13_plus=1 in14_plus=1 Name=Carol Age=14
in13_plus=1 in14_plus=1 Name=Henry Age=14
in13_plus=1 in14_plus=1 Name=Janet Age=15
in13_plus=1 in14_plus=0 Name=Jeffrey Age=13
in13_plus=1 in14_plus=1 Name=Judy Age=14
in13_plus=1 in14_plus=1 Name=Mary Age=15
in13_plus=1 in14_plus=1 Name=Philip Age=16
in13_plus=1 in14_plus=1 Name=Ronald Age=15
in13_plus=1 in14_plus=1 Name=William Age=15
```

Using PROC SQL

PROC SQL is a great tool for working with data tables and particularly for joining tables. However, it is not necessarily the best tool all the time, just as DATA step is not always the best tool. I think that PROC SQL is hard to debug. I also don't find it intuitive to write, because it isn't the top-down programming that I'm used to. But those issues notwithstanding, I still get a lot of use out of PROC SQL.

Choosing to Use PROC SQL

There are many ways to join data tables together. Should you use PROC SQL or should you use a DATA step with a merge? Unfortunately, the answer is, "It depends." I use both methods regularly, depending on the situation. Here are some of my reasons for choosing one method over the other.

Choose PROC SQL in the following situations:

- You are doing a simple join and you don't want to sort the data first.
- You are joining tables that don't have the same variables to merge by, or you want to use a more complex way of matching the variables in the tables (not just equals).
- You are joining more than two tables and you have different matching criteria for each join.

Choose a DATA step merge in the following situations:

- You want to keep records on the basis of which table they came from (using IN=).
- You need to do some additional DATA step processing, and you don't another step to read and write the data again after joining it.

Joining

Using PROC SQL to join tables is great because it lets you do the actual join in a much more complex way than a DATA step merge, which is a simple matching of variable values.

I tend to use the full join, left join, right join, and inner join rather than doing a Cartesian Product join (using a comma between the tables). I can get the same results, and I don't have to remember when to use a WHERE clause and when to use an ON clause. The cheat sheet in Figure 2.6 can help you to remember what each join does.

Figure 2.6: PROC SQL Join Cheat Sheet

Joins	Set Operators

Inner Join
-returns a single table with rows that have one or more matching rows in other table
-maximum of 16 tables at one time

```
select *
   from LEFT inner join RIGHT
      on LEFT.KEY = RIGHT.KEY
```

Left Join
-returns a single table with matching rows plus nonmatching rows from the left table
-can be performed on only 2 tables at a time

```
select *
   from LEFT left join RIGHT
      on LEFT.KEY = RIGHT.KEY
```

Right Join
-returns a single table with matching rows plus nonmatching rows from the right table
-can be performed on only 2 tables at a time

```
select *
   from LEFT right join RIGHT
      on LEFT.KEY = RIGHT.KEY
```

Full Join
-returns a single table with matching rows and nonmatching rows from both tables
-can be performed on only 2 tables at a time

```
select *
   from LEFT full join RIGHT
      on LEFT.KEY = RIGHT.KEY
```

Except
-unique rows from top table that are not in bottom table
-resulting columns determined by first table

```
select *
   from TOP
except
select *
   from BOTTOM
```

Intersect
-common, unique rows from both tables
-resulting columns determined by first table
-columns overlaid by order, not name

```
select *
   from TOP
intersect
select *
   from BOTTOM
```

Union
-all unique rows from both tables
-resulting columns determined by first table
-columns overlaid by order, not name

```
select *
   from TOP
union
select *
   from BOTTOM
```

Outer Union
-all rows from both tables, unique and non-unique
-all columns from both tables

```
select *
   from TOP
outer union
select *
   from BOTTOM
```

Using a Subquery

Using subqueries is a great feature of PROC SQL because it enables you to use the output of one query in an expression.

In the example in Program 2.11, if you want to find all the students who are older than the average age in the class, you can do it with a subquery. Figure 2.7 is the output table.

Program 2.11: Using a Subquery to Subset Data

```
proc sql;
   create table tallest as
      select name, sex, height
         from sashelp.class
            where height gt (select avg(height)
                                  from sashelp.class);
quit;
```

Figure 2.7: Using a Subquery to Subset Data

Obs	Name	Sex	Height
1	Alfred	M	69.0
2	Barbara	F	65.3
3	Carol	F	62.8
4	Henry	M	63.5
5	Janet	F	62.5
6	Jeffrey	M	62.5
7	Judy	F	64.3
8	Mary	F	66.5
9	Philip	M	72.0
10	Robert	M	64.8
11	Ronald	M	67.0
12	William	M	66.5

Another good use of a subquery is to compare a value to a set of values from a subquery using the IN operator. For example, the code in Program 2.12 enables you to keep all the students from the sashelp.class table who have the predict value in the sashelp.classfit table greater than 110. Figure 2.8 shows the output table.

Program 2.12: Using a Subquery with an IN Operator

```
proc sql;
   create table predict as
      select name, age, height
         from sashelp.class
            where name in (select name
                              from sashelp.classfit
                                 where predict gt 110);
quit;
```

Figure 2.8: Using a Subquery with an IN Operator

Obs	Name	Age	Height
1	Alfred	14	69.0
2	Barbara	13	65.3
3	Mary	15	66.5
4	Philip	16	72.0
5	Ronald	15	67.0
6	William	15	66.5

Using a Correlated Subquery

A correlated subquery is a subquery that refers to the parent query in the WHERE clause. This means that the subquery is run for each row of the parent query, so it is relatively inefficient.

Program 2.13 is an example of a correlated subquery that finds all the students who are taller than the average height by sex. So it compares each girl's height to the average of all the girls' heights, not to the average of all the heights. Figure 2.9 is a listing of the output table.

Program 2.13: Using a Correlated Subquery

```
proc sql;
   create table tallest as
      select name, sex, height
         from sashelp.class as c
            where height gt (select avg(height)
                                from sashelp.class
                                   where sex eq c.sex);
quit;
```

Figure 2.9: Using a Correlated Subquery

Obs	Name	Sex	Height
1	Alfred	M	69.0
2	Barbara	F	65.3
3	Carol	F	62.8
4	Janet	F	62.5
5	Judy	F	64.3
6	Mary	F	66.5
7	Philip	M	72.0
8	Robert	M	64.8
9	Ronald	M	67.0
10	William	M	66.5

Using Lookup Tables

Often in SAS you need to look up values from another table. For example, you might have a table that has userids in it, and you need to get the associated user name from a lookup table. There are lots of ways to do this based on what you are looking up and how big your tables are.

Using a Format

You can create a user-defined format to use if your lookup table meets the following criteria:

- It is a manageable size (you'll have to determine this by creating the format and seeing whether it takes too long to create and use or takes up too much disk space).
- It has only a couple of variables whose values you want to retrieve (you'll need a separate format for each variable).
- It has only one value to be retrieved for each lookup value.

The example in Program 2.14 and SAS Log 2.8 uses the sashelp.cars table and illustrates creating a lookup format of each car model to the make of car and a second format of car model to the type of car.

Program 2.14: Using Formats to Look Up Values

```
data fmtdata (keep = fmtname start label); ❶
   set sashelp.cars;
   start = left(model);

   fmtname = "$make";
   label = make;
   output;

   fmtname = "$type";
   label = type;
   output;
run;

proc sort data = fmtdata nodupkey;  ❷
   by fmtname start;
run;

proc format cntlin = fmtdata;    ❸
run;

data _null_;       ❹
   length model $50.;
   do model = "MDX", "ABC", "G500";
      type = put(model, $type.);
      make = put(model, $make.);
      putlog model= type= make=;
   end;
run;
```

❶ Create the table needed for PROC FORMAT. It must have a variable called fmtname that contains the name of the format, a variable called start that is the lookup variable's value, and a variable called label that contains the value to be retrieved.

❷ Group the data by fmtname if you are creating more than one format. You can have no duplicate lookup values (start). I handle both these requirements with a PROC SORT by fmtname and start and the NODUPKEY option.

❸ Run PROC FORMAT with the CNTLIN= option to create the format.

❹ Use a DATA step to test the formats. "ABC" is not a valid model, so it will not replace the model with the make and type.

SAS Log 2.8: Using Formats to Look Up Values

```
model=MDX type=MDX make=MDX
model=ABC type=ABC make=ABC
model=G500 type=G500 make=G500
```

See Also: There is a utility macro in Appendix A called %_MAKE_FORMAT() that you can use to create a format from a data table.

Using a Join

Another way to use a lookup table is to join your primary table with the lookup table. This can be done with either PROC SQL or with a DATA step and merge. You can use a join when your lookup table meets these criteria:

- It is any size.
- It has one or more values to be retrieved.
- It has one or more retrieval values per key (though this can be awkward because multiple records are created if there are multiple retrieval values).

The example in Program 2.15 and SAS Log 2.9 uses the sashelp.cars table as a lookup table and a straightforward DATA step to merge the data. Use the IN= option to keep only the records that come from your primary table.

Program 2.15: Using a DATA step Merge to Look Up Values

```
data primary;        ❶
    input model $char50.;
    datalines;
MDX
ABC
G500
;

data lookup;         ❷
    set sashelp.cars;
    model = left(model);
run;

proc sort data = primary;      ❸
    by model;
run;

proc sort data = lookup;
    by model;
run;

data _null_;       ❹
    merge primary (in = inP)
          lookup;
    by model;
    if (inP);
    putlog model= make= type=;
run;
```

❶ Create some test data for the primary table.
❷ Fix the cars data: Left-justify the model.
❸ Sort the data so that it can be merged.
❹ Merge the tables and keep only the records from the primary table using the IN= variable.

SAS Log 2.9: Using a DATA step Merge to Look Up Values

```
model=ABC Make=  Type=
model=G500 Make=Mercedes-Benz Type=SUV
model=MDX Make=Acura Type=SUV
```

Using PROC SQL is even easier than the DATA step, because you don't need to sort the tables. Also, in this example, you don't need to fix the model value in the lookup table. Program 2.16 uses a left join to ensure that you keep all the records from your primary table. The output table is listed in Figure 2.10.

Program 2.16: Using PROC SQL to Look Up Values

```
data primary;
   input model $char50.;
   datalines;
MDX
ABC
G500
;

proc sql;
   create table final as
       select p.model, l.make, l.type
           from primary as p
                left join
                sashelp.cars as l
                   on p.model eq left(l.model);
quit;
```

Figure 2.10: Using PROC SQL to Look Up Values

Obs	model	Make	Type
1	ABC		
2	G500	Mercedes-Benz	SUV
3	MDX	Acura	SUV

Using a SET Statement With KEY=

In a DATA step, you can also use a SET statement with a KEY= option to retrieve your values from the lookup table. To use this method, your lookup table must meet these criteria:

- It can be any size.
- It has one or more values to be retrieved.
- It has an index on the variable(s) that match the variable name(s) in the primary table.

The example in Program 2.17 and SAS Log 2.10 uses the sashelp.cars table as the lookup table and a SET statement with the KEY= option to look up values.

Program 2.17: Using SET with KEY= to Look Up Values

```
data primary;        ❶
   input model $char50.;
   datalines;
MDX
ABC
G500
;

data lookup (index = (model));        ❷
   set sashelp.cars;
   model = left(model);
run;

data _null_;        ❸
   length make $13 type $8;
   call missing(make, type);
   set primary;        ❹
   set lookup key = model;

   if (_iorc_ eq 0) then        ❺
      put "success: " model= make= type=;
   else
      put "failure: " model=;

   _error_ = 0;        ❻
run;
```

❶ Create some test data for the primary table.
❷ Fix the cars data: Left-justify the model. Add an index to the table.
❸ Run a DATA step to look up the values.
❹ Use one SET statement for the primary table and another SET statement with a KEY= option for the lookup table.
❺ Check the _IORC_ automatic variable to see if the value was found in the lookup table or not. A value of 0 means that it was found.
❻ Whenever you use a SET statement with a KEY= option, a MODIFY statement, or any other statements that cause the _IORC_ variable to be created, you should set the _ERROR_ automatic variable to 0. If you don't do this, the DATA step puts out all the variable values in the log every time the _IORC_ value is not 0.

SAS Log 2.10: Using SET with KEY= to Look Up Values

```
success: model=MDX make=Acura type=SUV
failure: model=ABC
success: model=G500 make=Mercedes-Benz type=SUV
```

If you have multiple records for each key value, you can use the same code and loop through the retrieved values instead of just reading one value. The code in Program 2.18 and SAS Log 2.11 shows how to do this.

Program 2.18: Using SET with KEY= to Look Up Multiple Values

```
data primary;      ❶
   input make $char50.;
   datalines;
XYZ
Mercedes-Benz
Audi
;

data lookup (index = (make));      ❷
   set sashelp.cars;
   model = left(model);
run;

data _null_;
   length model $40 type $8;
   call missing(model, type);
   set primary;        ❸

   reset = 1;
   do until(_iorc_ ne 0);      ❹

      set lookup key = make keyreset = reset;      ❺

      if (_iorc_ eq 0) then      ❻
         put make= model= type=;

   end; /* do until - loop though matching rows */

   _error_ = 0;      ❼
run;
```

❶ Create some test data for the primary table.
❷ Fix the cars data: Left-justify the model. Add an index to the table.
❸ Set the primary data table.
❹ Loop through the lookup table to retrieve all matches.
❺ Set the lookup table with the KEY= option pointing to the index. Use the KEYRESET= option to tell the SET statement to go to the next record in the table. Set the reset variable to 0 before the loop begins to tell the SET statement to read the first matching record.
❻ Whenever you use a SET statement with a KEY= option, a MODIFY statement, or any other statements that cause the _IORC_ variable to be created, you should set the _ERROR_ automatic variable to 0. If you don't do this, the DATA step puts out error messages in the log every time the _IORC_ value is not 0.

SAS Log 2.11: Using SET with KEY= to Look Up Multiple Values

```
make=Mercedes-Benz model=G500 type=SUV
make=Mercedes-Benz model=ML500 type=SUV
make=Mercedes-Benz model=C230 Sport 2dr type=Sedan
make=Mercedes-Benz model=C320 Sport 2dr type=Sedan
make=Mercedes-Benz model=C240 4dr type=Sedan
make=Mercedes-Benz model=C240 4dr type=Sedan
...
make=Audi model=A4 1.8T 4dr type=Sedan
```

```
make=Audi model=A41.8T convertible 2dr type=Sedan
make=Audi model=A4 3.0 4dr type=Sedan
make=Audi model=A4 3.0 Quattro 4dr manual type=Sedan
make=Audi model=A4 3.0 Quattro 4dr auto type=Sedan
make=Audi model=A6 3.0 4dr type=Sedan
...
```

Using Hash Tables

Using a hash table in a DATA step is an excellent option if your lookup table is very large. Setting up the lookup table as a hash table puts the data into memory so that the values can be retrieved much more quickly. If the neither of your tables is very large, then using a hash table is probably less efficient than some of the other lookup methods.

Program 2.19 and SAS Log 2.12 show the same example again with sashelp.cars as the lookup table. This example shows how to create and use a basic hash table. Note that there are lots of things you can do with a hash table. This is one of the more common uses.

Program 2.19: Using a Hash Table to Look Up Values

```
data primary;      ❶
   input model $char40.;
   datalines;
MDX
ABC
G500
;

data lookup;      ❷
   set sashelp.cars;
   model = left(model);
run;

data _null_;

   if 0 then
      set lookup;      ❸

   if _n_=1  then
   do;
      declare hash H (dataset: "lookup");   ❹
      rc = H.defineKey("model");
      rc = H.defineData("make", "type");
      rc = H.defineDone();
      call missing(make, type);   ❺
   end;

   set primary;      ❻

   rc = H.find();      ❼
   if rc eq 0 then       ❽
      putlog "success: " model= make= type=;
   else
      putlog "failure: " model=;

run;
```

❶ Create some test data for the primary table.

❷ Fix the cars data: Left-justify the model.

❸ The hash table variables must be defined in the DATA step before you create the hash table. I use the if (0) syntax to bring all the variable definitions from the lookup table into the DATA step without having to define them all.

❹ Create the hash table with a DECLARE statement. Then define the key and the data.

❺ Set all the hash data values to missing so that you don't get notes about uninitialized variables.

❻ Start reading from the primary table.

❼ Search the hash table for a matching key value.

❽ If the return code from the FIND function is 0, then a match was found.

SAS Log 2.12: Using a Hash Table to Look Up Values

```
success: Model=MDX Make=Acura Type=SUV
failure: Model=ABC
success: Model=G500 Make=Mercedes-Benz Type=SUV
```

Updating Data In Place

There are times that you need to update a data table *in place*, meaning that you just want to update some of the records in a table without rewriting the table completely. This is useful when the table is very big, when you don't want to lose any of the attributes of the original table, or when you are updating a DBMS table. There are several ways to do so, including PROC SQL, a DATA step with a MODIFY statement, and PROC APPEND.

For the following examples, I used the code in Program 2.20 to make a copy of the sashelp.class table in the work library, since I don't want to permanently change the sashelp table:

Program 2.20: Creating Test Data

```
proc copy in = sashelp out = work;
   select class;
run;
```

Using PROC SQL

There are three different statements in PROC SQL that you can use to update a table in place: INSERT, DELETE and UPDATE.

Using the INSERT Clause

The INSERT clause adds records to the data table using the VALUES, SET, or SELECT clauses. Each of the following examples insert two new rows into the class table.

The example in Program 2.21 and SAS Log 2.13 uses the VALUES clause, which lets you list the values that will be inserted into the table. The values must be in the same order as the variables in the table.

Program 2.21: Using INSERT with VALUES Clauses

```
proc sql;
   insert into class
      values("Martha", "F", 14, 64, 106)
      values("Chris", "M", 13, 67, 111);
quit;
```

SAS Log 2.13: Using INSERT with VALUES Clauses

```
NOTE: 2 rows were inserted into WORK.CLASS.
```

The example in Program 2.22 and SAS Log 2.14 uses the SET clause to specify each variable and its value.

Program 2.22: Using INSERT with SET Clauses

```
proc sql;
   insert into class
      set name = "Martha", sex = "F", age = 14,
         height = 64, weight = 106
      set name = "Chris", sex = "M", age = 13,
         height = 67, weight = 111;
quit;
```

SAS Log 2.14: Using INSERT with SET Clauses

```
NOTE: 2 rows were inserted into WORK.CLASS.
```

The example in Program 2.23 and SAS Log 2.15 shows inserting rows from another table using a SELECT/FROM clause. The variables in the second table must be in the same order as the variables in the main table.

Program 2.23: Using INSERT with a SELECT Clause

```
data second;
   if (0) then
      set class;
   input name sex age height weight;
   datalines;
Martha F 14 64 106
Chris M 13 67 111
;

proc sql;
   insert into class
      select *
         from second;
quit;
```

SAS Log 2.15: Using INSERT with a SELECT Clause

```
NOTE: 2 rows were inserted into WORK.CLASS.
```

Using the DELETE Clause

The DELETE clause enables you to delete records from a table.

The example in Program 2.24 and SAS Log 2.16 illustrates deleting all the records from the table where the value of name is "Jane".

Program 2.24: Deleting Records with a WHERE Clause

```
proc sql;
   delete from class
      where name eq "Jane";
quit;
```

SAS Log 2.16: Deleting Records with a WHERE Clause

```
NOTE: 1 row was deleted from WORK.CLASS.
```

The example in Program 2.25 and SAS Log 2.17 illustrates deleting all the records in the table, so be careful!

Program 2.25: Deleting All Records

```
proc sql;
   delete from class;
quit;
```

SAS Log 2.17: Deleting All Records

```
NOTE: 22 rows were deleted from WORK.CLASS.
```

Using the UPDATE Clause

The UPDATE clause updates the values for the given variables. It can update all the records or just the records that match a WHERE condition.

The example in Program 2.26 and SAS Log 2.18 sets the value of age to 15 for all the records where the value of name is "Jane".

Program 2.26: Updating Records

```
proc sql;
   update class
      set age = 15
         where name eq "Jane";
quit;
```

SAS Log 2.18: Updating Records

```
NOTE: 1 row was updated in WORK.CLASS.
```

Using the MODIFY Statement

Using a DATA step with the MODIFY statement is quite powerful because you can use programming statements to conditionally insert, delete and update records and values in a single step.

You need to use the _IORC_ automatic variable, which is created for you by the MODIFY statement, and the %SYSRC() macro to check the return code, as follows:

- The value of _IORC_ is 0 if the record was found in the main table.
- The value of _IORC_ is %SYSRC(_DSENMR) if the record was not found in the main table.

If the value of _IORC_ is not 0, then the DATA step puts all the variable values into the SAS log for this record. You can stop this information from filling up the SAS log by setting the _ERROR_ variable to 0.

The example in Program 2.27 and SAS Log 2.19 combines the main class table with a second table using a BY statement. The program performs the following actions:

- If the name is in both tables, and the sex is M, then the record is deleted from the class table.
- If the name is in both tables, and the sex is F, then the record is updated with the values from the second table.
- If the name is not in the class table, then the record from the second table is added to the class table.

Program 2.27: Updating with MODIFY and BY Statements

```
data second;
   if (0) then
      set class;
   input name sex age height weight;
   datalines;
Jane F 14 64 106
Robert M 15 69 113
Mark M 13 67 111
;
```

```
data class;
   modify class second;
   by name;
   if (_iorc_ eq 0) then
   do;
      if (sex eq "M") then
         remove;
      else
         replace;
   end;
   else if (_iorc_ eq %sysrc(_dsenmr)) then
      output;
   _error_ = 0;
run;
```

SAS Log 2.19: Updating with MODIFY and BY Statements

```
NOTE: The data set WORK.CLASS has been updated. There were 1
observations rewritten, 1 observations added and 1 observations
deleted.
```

The modify statement can also be used when the main table is indexed so that you look for a key variable instead of using a BY statement. The example in Program 2.28 and SAS Log 2.20 shows the same process as the previous one, but uses an index on the class table. Note that you do need to check for a different _IORC_ return code, %SYSRC(_DSENOM), when looking for non-matches.

Program 2.28: Updating with MODIFY and the KEY= Option

```
data second;
   if (0) then
      set class;
   input name sex age height weight;
   datalines;
Jane F 14 64 106
Robert M 15 69 113
Mark M 13 67 111
;

proc datasets lib = work nolist nowarn;
   modify class;
      index create name;
quit;
```

```
data class;
   set second;
   modify class key = name;
   if (_iorc_ eq 0) then
   do;
      if (sex eq "M") then
         remove;
      else
         replace;
   end;
   else if (_iorc_ eq %sysrc(_dsenom)) then
      output;
   _error_ = 0;
run;
```

SAS Log 2.20: Updating with MODIFY and the KEY= Option

```
NOTE: The data set WORK.CLASS has been updated. There were 1
observations rewritten, 1 observations added and 1 observations
deleted.
```

Using PROC APPEND

You can also use PROC APPEND to update a table in place. It does not read or write the records that are in the main table (BASE=); it just adds the records from the second data table (DATA=) to the end of the main table. It is much more efficient than concatenating tables using a DATA step with a SET statement, because the first table does not need to be read and written. You do need to be sure that the second table has the same structure as the first table—the same variables with the same attributes. You get error messages unless you use the FORCE option, and even with FORCE, you still get warning messages if the lengths are different.

The example in Program 2.29 and SAS Log 2.21 uses the technique described in the section "Copying Variable Attributes" to get the correct variable attributes for the second table.

Program 2.29: Updating with PROC APPEND

```
data second;
   if (0) then
      set class;
   input name sex age height weight;
   datalines;
Martha F 14 64 106
Chris M 13 67 111
;

proc append base = class data = second;
run;
```

SAS Log 2.21: Updating with PROC APPEND

```
NOTE: Appending WORK.SECOND to WORK.CLASS.
NOTE: There were 2 observations read from the data set WORK.SECOND.
NOTE: 2 observations added.
```

Finding Records

"Does this table have any records?", "How many records are in this table?", and the related "Does this table exist?" are all very common questions that need to be answered in SAS programs.

See Also: In Appendix A there is a utility macro called %_GET_NUM_RECORDS that you can use to answer all of these questions with a single macro call.

Determining Whether a Table Exists

To check to see whether a data table exists, you can use the EXIST() function in a DATA step or in macro code with %SYSFUNC(). The function returns 1 (true) if the table exists and 0 (false) if the table does not exist.

The example in Program 2.30 and SAS Log 2.22 shows how to use the EXIST() function in a DATA step.

Program 2.30: Checking for Existence in a DATA Step

```
data _null_;
   do table = "sashelp.class", "sashelp.xyz";
      if (exist(table)) then
         putlog table "exists";
      else
         putlog table "does not exist";
   end;
run;
```

SAS Log 2.22: Checking for Existence in a DATA Step

```
sashelp.class exists
sashelp.xyz does not exist
```

In a macro, you can use the EXIST() function with the %SYSFUNC() macro function. Program 2.31 and SAS Log 2.23 show how to use the EXIST() function in a macro with the %SYSFUNC() macro function.

Program 2.31: Checking for Existence in a Macro

```
%macro testExist(table);

   %if (%sysfunc(exist(&table))) %then
      %put &table exists;
   %else
      %put &table does not exist;

%mend testExist;
%testExist(sashelp.class);
%testExist(sashelp.xyz);
```

SAS Log 2.23: Checking for Existence in a Macro

```
11   %testExist(sashelp.class);
sashelp.class exists
12   %testExist(sashelp.xyz);
sashelp.xyz does not exist
```

Getting the Number of Records

Several different methods enable you to get the number of records in a table. Some of these methods work better than others, but all are valid.

One of the things to keep in mind when you are looking at the number of records in a SAS data table is that tables can contain deleted records. You can't see the deleted records, but you can see missing rows when you look at the OBS number that is displayed in a PROC PRINT or in the VIEWTABLE window. Deleted records occur when you use a method that deletes records without rewriting the entire table.

Using Brute Force

Just count the records. This works great if you are already reading the data table. If you are not already reading the data table, this is not very efficient. You can either use the _N_ variable, or you can create a counter. Program 2.32 and SAS Log 2.24 show both methods.

Program 2.32: Using Brute Force to Get the Number of Records

```
%let numRecords = 0;

data _null_;
   set sashelp.class end = eof;
   count + 1;
   if (eof) then
   do;
       call symputx("numRecords", count);
       call symputx("numRecords2", _n_);
   end;
run;

%put &=numRecords;
%put &=numRecords2;
```

SAS Log 2.24: Using Brute Force to Get the Number of Records

```
12   %put &=numRecords;
NUMRECORDS=19
13   %put &=numRecords2;
NUMRECORDS2=19
```

Tip: If you are setting macro variable values in a DATA step, be sure to initialize the variables before the DATA step. If there are no records in the table that you are reading, then the DATA step code doesn't execute, and the macro variables are not created. You then get a warning/error if you attempt to use the macro variable later in your program.

Using the NOBS= Option

This option is better than the brute force option, but is not very elegant, still a little clunky. However, it might make sense in certain situations. The NOBS= option on a SET statement creates a variable that contains the number of records before the DATA step even starts, so it isn't necessary to physically read any records from the table. Program 2.33 and SAS Log 2.25 show how to do this.

Program 2.33: Using the NOBS= Option to Get the Number of Records

```
%let numRecords = 0;

data _null_;
   if (0) then
      set sashelp.class nobs = numRecords;
   call symputx("numRecords", numRecords);
   stop;
run;

%put &=numRecords;
```

SAS Log 2.25: Using the NOBS= Option to Get the Number of Records

```
10  %put &=numRecords;
NUMRECORDS=19
```

Using PROC CONTENTS

There is a lot of great information available in the output from PROC CONTENTS, including the number of records in the table. The OUT= data table contains a record for each variable in the table, and the table-level information (including the number of records) is repeated on each record of the OUT= data table. The NOBS variable in the OUT= data table contains the number of records in the table.

I don't recommend this method unless you need the PROC CONTENTS output for another purpose. It requires a procedure and a DATA step and creates a temporary table. That is a lot of work just to get the number of records.

Tip: When using PROC CONTENTS, if you are creating the OUT= data table, be sure to include the NOPRINT option on the PROC CONTENTS statement to avoid creating unnecessary reports.

Program 2.34 and SAS Log 2.26 show how to use PROC CONTENTS to get the number of records.

Program 2.34: Using PROC CONTENTS to Get the Number of Records

```
%let numRecords = 0;

proc contents data = sashelp.class
              out = contents
              noprint;
run;

data _null_;
   set contents(obs = 1);
   call symputx("numRecords", nobs);
run;

%put &=numRecords;
```

SAS Log 2.26: Using PROC CONTENTS to Get the Number of Records

```
1    %let numRecords = 0;
2
3    proc contents data = sashelp.class
4                  out = contents
5                  noprint;
6    run;

NOTE: The data set WORK.CONTENTS has 5 observations and 41
variables.
NOTE: PROCEDURE CONTENTS used (Total process time):
      real time           0.02 seconds
      cpu time            0.01 seconds

7
8    data _null_;
9        set contents(obs = 1);
10       call symputx("numRecords", nobs - delobs);
11   run;

NOTE: There were 1 observations read from the data set
WORK.CONTENTS.
NOTE: DATA statement used (Total process time):
      real time           0.00 seconds
      cpu time            0.00 seconds

12
13   %put &=numRecords;
NUMRECORDS=19
```

Using the sashelp.vtable View

The sashelp.vtable view contains most of the header information for all tables that are defined in your SAS session. There is one record in the view for each table in the SAS session, so you must subset the view to locate the table that you want (using the LIBNAME and MEMNAME variables), and then get the number of records from the NLOBS variable. Note that using this view might not be very efficient if you have a lot of tables defined in your SAS session.

Program 2.35 and SAS Log 2.27 show how to use the sashelp.vtable to get the number of records in the sashelp.class table.

Program 2.35: Using sashelp.vtable to Get the Number of Records

```
%let numRecords = 0;

data _null_;
   set sashelp.vtable;
   where libname = "SASHELP" and memname = "CLASS";
   call symputx("numRecords", nlobs);
run;

%put &=numRecords;
```

SAS Log 2.27: Using sashelp.vtable to Get the Number of Records

```
1    %let numRecords = 0;
2
3    data _null_;
4       set sashelp.vtable;
5       where libname = "SASHELP" and memname = "CLASS";
6       call symputx("numRecords", nlobs);
7    run;

NOTE: There were 1 observations read from the data set
SASHELP.VTABLE.
      WHERE (libname='SASHELP') and (memname='CLASS');
NOTE: DATA statement used (Total process time):
      real time           0.13 seconds
      cpu time            0.12 seconds

8
9    %put &=numRecords;
NUMRECORDS=19
```

Using the ATTRN() Function

The ATTRN() function can be used to get lots of information about a data table, and usually it does not require reading any records from the table. For the number of records, you can look at various attributes:

NOBS
: The total number of records, including deleted records.

NLOBS
: The total number of undeleted records.

NLOBSF

The total number of undeleted records with a WHERE clause, or with the OBS= or FIRSTOBS= options applied.

To calculate this value, the function must read all the records in the table. The NOBS and NLOBS attributes are saved in the data table's header, so the function doesn't need to read all the records for those attributes.

I would suggest using NLOBS most of the time unless you want to see the number of records in a subset. In that case, use NLOBSF.

Some DBMS tables might not provide the information that you can get from a SAS table (including the number of records). In that case, you need to use the brute force method described earlier.

Program 2.36 and SAS Log 2.28 show how to use the ATTRN() function in a DATA step.

Program 2.36: Using ATTRN() in a DATA Step to Get the Number of Records

```
%let numRecords = 0;

data _null_;
    table = "sashelp.class";
    dsid = open(table);
    if (dsid gt 0) then
    do;
        nobs = attrn(dsid, "NLOBS");
        call symputx("numRecords", nobs);
        dsid = close(dsid);
    end;
run;

%put &=numRecords;
```

SAS Log 2.28: Using ATTRN() in a DATA Step to Get the Number of Records

```
14   %put &=numRecords;
NUMRECORDS=19
```

Program 2.37 and SAS Log 2.29 show how to use ATTRN() with %SYSFUNC() in a macro.

Program 2.37: Using ATTRN() in a Macro to Get the Number of Records

```
%macro numRecords(table);

    %let dsid = %sysfunc(open(&table));
    %if (&dsid gt 0) %then
    %do;
        %let numRecords = %sysfunc(attrn(&dsid, NLOBS));
        %let dsid = %sysfunc(close(&dsid));
    %end;
```

```
%mend numRecords;

%let numRecords = 0;
%numRecords(sashelp.class);
%put &=numRecords;
```

SAS Log 2.29: Using ATTRN() in a Macro to Get the Number of Records

```
14   %put &=numRecords;
NUMRECORDS=19
```

Verifying That a Table Has Records

If you find out how many records are in the table, you automatically know whether there are any records, because the number is greater than 0. A few other methods can tell you whether you have any records.

Using Brute Force

The brute force method for finding out whether a table has any records involves trying to read the table in a DATA step. If there are no records, then none of the code in the DATA step runs.

The first DATA step in the example in Program 2.38 and SAS Log 2.30 checks to see whether there are any records in the sashelp.class table. The second DATA step checks the same table, this time with a WHERE clause applied, which returns no records. This method relies on setting the &ANYRECORDS macro variable to 0 before running the DATA step. If there are no records in the data table, then the DATA step doesn't execute, and &ANYRECORDS keeps the value of 0.

Program 2.38: Using Brute Force to See Whether There Are Any Records

```
%let anyRecords = 0;

data _null_;
   set sashelp.class;
   call symput("anyRecords", "1");
   stop;
run;

%put &=anyRecords;

%let anyRecords = 0;

data _null_;
   set sashelp.class;
   where name eq "Martha";
   call symput("anyRecords", "1");
   stop;
run;

%put &=anyRecords;
```

SAS Log 2.30: Using Brute Force to See Whether There Are Any Records

```
9   %put &=anyRecords;
ANYRECORDS=1
...
```

```
20   %put &=anyRecords;
ANYRECORDS=0
```

Using the ATTRN() Function

The ATTRN() function has an ANY attribute that you can query. The ANY attribute has a value of 0 when there are no records in the table, -1 if there are no records or no variables in the table, and 1 if there are records in the table.

Program 2.39 and SAS Log 2.31 show how to use the ATTRN() function in a DATA step to query a table for any records.

Program 2.39: Using ATTRN() in a DATA Step to See Whether There Are Any Records

```
data oneRecordOneVariable;
    x=1;
run;
data noRecordsOneVariable;
    set oneRecordOneVariable;
    stop;
run;
data noRecordsNoVariables;
    stop;
run;

data _null_;
    do table = "oneRecordOneVariable",
               "noRecordsOneVariable",
               "noRecordsNoVariables",
               "xxx";
        dsid = open(table);
        if (dsid gt 0) then
        do;
            any = attrn(dsid, "ANY");
            if (any gt 0) then
                putlog "There are records in " table;
            else if (any eq 0) then
                putlog "There are no records in " table;
            else
                putlog "There are no records/variables in " table;
            dsid = close(dsid);
        end;
        else
            putlog table " could not be opened";
    end;
run;
```

SAS Log 2.31: Using ATTRN() in a DATA step to See Whether There Are Any Records

```
There are records in oneRecordOneVariable
NOTE: No observations in data set WORK.NORECORDSONEVARIABLE.
There are no records in noRecordsOneVariable
NOTE: No variables in data set WORK.NORECORDSNOVARIABLES.
There are no records/variables in noRecordsNoVariables
xxx   could not be opened
```

Program 2.40 and SAS Log 2.32 show that this can be done in a similar way with a macro.

Program 2.40: Using ATTRN() in a Macro to See Whether There Are Any Records

```
data oneRecordOneVariable;
    x=1;
run;
data noRecordsOneVariable;
    set oneRecordOneVariable;
    stop;
run;
data noRecordsNoVariables;
    stop;
run;

%macro anyRecords(table);

    %let dsid = %sysfunc(open(&table));
    %if (&dsid gt 0) %then
    %do;
        %let any = %sysfunc(attrn(&dsid, ANY));
        %if (&any gt 0) %then
            %put There are records in &table;
        %else %if (&any eq 0) %then
            %put There are no records in &table;
        %else
            %put There are no records/variables in &table;
        %let dsid = %sysfunc(close(&dsid));
    %end;
    %else
        %put &table could not be opened;

%mend anyRecords;

%anyRecords(oneRecordOneVariable);
%anyRecords(noRecordsOneVariable);
%anyRecords(noRecordsNoVariables);
%anyRecords(xxx);
```

SAS Log 2.32: Using ATTRN() in a Macro to See Whether There Are Any Records

```
31  %anyRecords(oneRecordOneVariable);
There are records in oneRecordOneVariable
32  %anyRecords(noRecordsOneVariable);
NOTE: No observations in data set WORK.NORECORDSONEVARIABLE.
There are no records in noRecordsOneVariable
33  %anyRecords(noRecordsNoVariables);
NOTE: No variables in data set WORK.NORECORDSNOVARIABLES.
There are no records/variables in noRecordsNoVariables
34  %anyRecords(xxx);
xxx could not be opened
```

Re-creating Indexes

Sometimes it is necessary to rewrite a data table that has an index on it, so you need to re-create the index on the updated table. But what if you don't know what the index was? PROC CONTENTS has a second output table that can be created using the OUT2= option. The OUT2 table contains information about the indexes, including a great variable called RECREATE that has the actual PROC DATASETS code for creating the index. So as long as you run PROC CONTENTS before you modify the table, you can then re-create the index on the updated table. The RECREATE variable also contains the statement to re-create any integrity constraints that might be on the table. The example in Program 2.41 and SAS Log 2.33 shows how to use the RECREATE variable to add an index to the cars table that was retrieved from the cars_with_index table.

Program 2.41: Re-creating an Index

```
data cars_with_index (index = (make_type = (make type) model))
      cars;
   set sashelp.cars;
run;

proc contents data = work.cars_with_index
              out2 = indexes
              noprint;
run;

data _null_;
   set indexes end = eof;
   if (_n_ eq 1) then
   do;
      call execute("proc datasets lib = work nowarn nolist;");
      call execute("   modify cars;");
   end;
   call execute("      " !! recreate);
   if (eof) then
      call execute("quit;");
run;
```

SAS Log 2.33: Re-creating an Index

```
1    data cars_with_index (index = (make_type = (make type) model))
2          cars;
3       set sashelp.cars;
4    run;

NOTE: There were 428 observations read from the data set
SASHELP.CARS.
NOTE: The data set WORK.CARS_WITH_INDEX has 428 observations and 15
variables.
NOTE: The data set WORK.CARS has 428 observations and 15 variables.
NOTE: DATA statement used (Total process time):
      real time            0.03 seconds
      cpu time             0.01 seconds

5
6    proc contents data = work.cars_with_index
7                  out2 = indexes
```

```
8                     noprint;
9    run;

NOTE: The data set WORK.INDEXES has 2 observations and 21 variables.
NOTE: PROCEDURE CONTENTS used (Total process time):
      real time           0.01 seconds
      cpu time            0.00 seconds


10
11   data _null_;
12      set indexes end = eof;
13      if (_n_ eq 1) then
14      do;
15         call execute("proc datasets lib = work nowarn nolist;");
16         call execute("   modify cars;");
17      end;
18      call execute("       " !! recreate);
19      if (eof) then
20         call execute("quit;");
21   run;

NOTE: There were 2 observations read from the data set WORK.INDEXES.
NOTE: DATA statement used (Total process time):
      real time           0.00 seconds
      cpu time            0.01 seconds

NOTE: CALL EXECUTE generated line.
1     + proc datasets lib = work nowarn nolist;
2     +    modify cars;
3     +       Index create Model / Updatecentiles=5;
NOTE: Simple index Model has been defined.
4     +       Index create make_type=( Make Type ) /
Updatecentiles=5;
NOTE: Composite index make_type has been defined.
5     + quit;

NOTE: MODIFY was successful for WORK.CARS.DATA.
NOTE: PROCEDURE DATASETS used (Total process time):
```

See Also: This example uses CALL EXECUTE() to generate the SAS code that creates the index. CALL EXECUTE() is a great tool that is discussed in Chapter 4 in the section "Creating and Running Code Based on Data Values."

Chapter 3: The Operating System

Introduction

In a SAS program, you sometimes need to access the operating system (OS) to run system commands and to access the file system. Reading data from external files is usually a requirement in a SAS application; you have to get the data into tables before you can analyze it. This chapter contains some tips and techniques that will help you navigate the world beyond SAS. Since Windows and Unix are the most popular operating systems with SAS programmers, I am going to focus only on those environments. Many of these programs work on other operating systems; see the relevant *SAS Companion* guide to work with their idiosyncrasies.

Checking the Operating System

To work with the operating system, you must know what OS the program is running on, especially if you are going to run the same program on multiple operating systems. Table 3.1 shows the automatic macro variables &SYSSCP and &SYSSCPL and their values on the different operating systems.

Table 3.1: The Values of &SYSSCP and &SYSSCPL

Operating System	&SYSSCP	&SYSSCPL
Windows	WIN	Version of Windows
Unix	Unix processor (for example, LIN X64, SUN 64, AIX 64)	Specific Unix environment (for example, HP-UX, SunOS, AIX)

If I am writing a program that might run on any of these systems, I generally check whether the value of &SYSSCP is WIN, and I use an open-ended ELSE for Unix (because the values vary on the basis of the specific Unix processor and environment). You can see an example in Program 3.1. SAS Log 3.1 shows the log when the program was run on Windows (64-bit Windows 7 Pro), and SAS Log 3.2 shows the log when the program was run on Unix (64-bit Linux).

Program 3.1: Checking the Operating System

```
%macro check_os;
   %if (&sysscp eq WIN) %then
      %put Windows: &sysscpl;
   %else %if (&sysscp eq OS) %then
      %put z/OS: &sysscpl;
   %else
      %put Unix: &sysscp &sysscpl;
%mend check_os;
%check_os;
```

SAS Log 3.1: Checking the Operating System - Windows

```
Windows: X64_7PRO
```

SAS Log 3.2: Checking the Operating System - Unix

```
Unix: LIN X64 Linux
```

Running Operating System Commands

There are multiple ways to run an operating system command from SAS.

To run an OS command from a SAS program, you must have the XCMD system option turned on. If you are running SAS in batch or interactively using Display Manager or line mode, XCMD is usually turned on by default. If it is not turned on, you can add it to the SAS config file, or you can specify it on the sas command.

If you are running on a workspace server (for example, from a stored process or from SAS Enterprise Guide), this option is usually set to NOXCMD. Because this is a configuration or invocation option only (it must be specified as SAS is started), the best place to set this option is in SAS Management Console, as follows:

1. On the **Plug-ins** tab, expand the **Server Manager**.
2. Expand the App Server that you are using (SASApp is the most common one).

3. Expand the **Logical Workspace Server**.
4. Right-click the **Workspace Server** and choose **Properties**.
5. Go to the **Options** tab.
6. Click the **Advanced Options** button.
7. Go to the **Launch Properties** tab.
8. Select the **Allow XCMD** option.
9. The workspace server will need to be restarted after setting this option, but once it is set, it will be set for all sessions.

Tip—Windows: You might need to specify the (NO)XWAIT and (NO)XSYNC system options. To return processing to your SAS session when the command finishes, you must have NOXWAIT set. Otherwise, you need to close the Command window manually in order to get your SAS program to continue running. Setting XSYNC makes the OS command finish processing before control returns to SAS. Setting NOXSYNC lets the OS command run in the background while your SAS program continues processing.

Tip—Windows: You can execute multiple commands in a single statement by using an ampersand (&) between the commands: cd c:\temp & mkdir temp2

Tip—Unix: You can execute multiple commands in a single statement by using a semicolon (;) between the commands: cd /tmp; mkdir temp2

Using the X Statement

The X statement and the %SYSEXEC macro statement can be executed from anywhere in a SAS program to run an OS command:

```
x "mkdir newFolder";

%sysexec "mkdir newFolder";
```

This is the most efficient method of running a command and can be conditionally executed by macro statements if necessary. The return code from the command is saved in the &SYSRC automatic macro variable: 0 = success, >0 = failure. If you execute multiple commands in a single X statement, you get a negative value in the &SYSRC.

Using the SYSTEM() Function

The SYSTEM() function and the CALL SYSTEM() routine can be used in a DATA step. These are useful if you want to create commands based on the contents of a data table. The SYSTEM() function produces a return code that is 0 for success and >0 for failure. A negative value generally indicates a problem, but it depends on the operating system and the command. The CALL SYSTEM() routine does not produce a retrievable return code, which is why I never use it.

The example in Program 3.2 creates a directory for each person in the sashelp.class table. It also determines which directory to use based on the operating system. Figure 3.1 shows the results on Windows and Figure 3.2 shows the results on Unix.

Program 3.2: Using the SYSTEM() Function to Create Directories

```
options noxwait;

data _null_;
   set sashelp.class;
   if (_n_ eq 1) then
   do;
      if ("&sysscp" eq "WIN") then
         rc = system("cd c:/temp");
      else
         rc = system("cd /tmp");
      if (rc ne 0) then
      do;
         putlog "ERR" "OR: Unable to get to $HOME";
         stop;
      end;
      rc = system("mkdir class");
      rc = system("cd class");
   end;
   call system("mkdir " !! name);
run;
```

Figure 3.1: Using the SYSTEM() Function to Create Directories—Windows

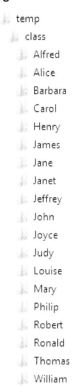

- temp
- class
- Alfred
- Alice
- Barbara
- Carol
- Henry
- James
- Jane
- Janet
- Jeffrey
- John
- Joyce
- Judy
- Louise
- Mary
- Philip
- Robert
- Ronald
- Thomas
- William

Figure 3.2: Using the SYSTEM Function to Create Directories—Unix

```
$ cd /tmp
$ cd class
$ ls
Alfred  Barbara  Henry   Jane    Jeffrey  Joyce  Louise  Philip  Ronald  William
Alice   Carol    James   Janet   John     Judy   Mary    Robert  Thomas
```

Using a FILENAME Pipe

Another way to run an OS command is to use a FILENAME statement with a pipe device, and a DATA step with INFILE and INPUT statements. I use this method quite often because it gives the most control. You can retrieve the results from the command, including error messages.

The code in Program 3.3 works on Windows and Unix to give you a directory listing. The output from the command is read with the DATA step and written back into the log. SAS Log 3.3 was created on Windows, and SAS Log 3.4 was created on Unix.

Program 3.3: Using a FILENAME Pipe to Get a Directory Listing

```
%macro runPipe;

   %if (&sysscp eq WIN) %then
   %do;
      filename os_cmd pipe "cd c:\temp\class & dir";
   %end;
   %else
   %do;
      filename os_cmd pipe "cd /tmp/class; ls -l";
   %end;

   data _null_;
      infile os_cmd;
      input;
      putlog _infile_;
   run;

   filename os_cmd;

%mend;
%runPipe
```

SAS Log 3.3: Using a FILENAME Pipe to Get a Directory Listing—Windows

```
NOTE: The infile OS_CMD is:
      Unnamed Pipe Access Device,
      PROCESS=cd c:\temp\class & dir,RECFM=V,
      LRECL=32767

 Volume in drive C is Win7x64
 Volume Serial Number is 7217-679F

 Directory of c:\temp\class

01/01/2017 10:30 AM  <DIR>     .
01/01/2017 10:30 AM  <DIR>     ..
01/01/2017 10:30 AM  <DIR>     Alfred
```

```
01/01/2017 10:30 AM  <DIR>     Alice
01/01/2017 10:30 AM  <DIR>     Barbara
01/01/2017 10:30 AM  <DIR>     Carol
01/01/2017 10:30 AM  <DIR>     Henry
01/01/2017 10:30 AM  <DIR>     James
01/01/2017 10:30 AM  <DIR>     Jane
01/01/2017 10:30 AM  <DIR>     Janet
01/01/2017 10:30 AM  <DIR>     Jeffrey
01/01/2017 10:30 AM  <DIR>     John
01/01/2017 10:30 AM  <DIR>     Joyce
01/01/2017 10:30 AM  <DIR>     Judy
01/01/2017 10:30 AM  <DIR>     Louise
01/01/2017 10:30 AM  <DIR>     Mary
01/01/2017 10:30 AM  <DIR>     Philip
01/01/2017 10:30 AM  <DIR>     Robert
01/01/2017 10:30 AM  <DIR>     Ronald
01/01/2017 10:30 AM  <DIR>     Thomas
04/13/2017 10:30 AM  <DIR>     William
         0 File(s)        0 bytes
        21 Dir(s)
NOTE: 28 records were read from the infile OS_CMD.
      The minimum record length was 0.
      The maximum record length was 50.
```

SAS Log 3.4: Using a FILENAME pipe to Get a Directory Listing—Unix

```
NOTE: The infile OS_CMD is:
      Pipe command="cd /tmp/class; ls -l"

total 76
drwxr-xr-x. 2 userid group 4096 Jan 1 10:30 Alfred
drwxr-xr-x. 2 userid group 4096 Jan 1 10:30 Alice
drwxr-xr-x. 2 userid group 4096 Jan 1 10:30 Barbara
drwxr-xr-x. 2 userid group 4096 Jan 1 10:30 Carol
drwxr-xr-x. 2 userid group 4096 Jan 1 10:30 Henry
drwxr-xr-x. 2 userid group 4096 Jan 1 10:30 James
drwxr-xr-x. 2 userid group 4096 Jan 1 10:30 Jane
drwxr-xr-x. 2 userid group 4096 Jan 1 10:30 Janet
drwxr-xr-x. 2 userid group 4096 Jan 1 10:30 Jeffrey
drwxr-xr-x. 2 userid group 4096 Jan 1 10:30 John
drwxr-xr-x. 2 userid group 4096 Jan 1 10:30 Joyce
drwxr-xr-x. 2 userid group 4096 Jan 1 10:30 Judy
drwxr-xr-x. 2 userid group 4096 Jan 1 10:30 Louise
drwxr-xr-x. 2 userid group 4096 Jan 1 10:30 Mary
drwxr-xr-x. 2 userid group 4096 Jan 1 10:30 Philip
drwxr-xr-x. 2 userid group 4096 Jan 1 10:30 Robert
drwxr-xr-x. 2 userid group 4096 Jan 1 10:30 Ronald
drwxr-xr-x. 2 userid group 4096 Jan 1 10:30 Thomas
drwxr-xr-x. 2 userid group 4096 Jan 1 10:30 William
NOTE: 20 records were read from the infile OS_CMD.
      The minimum record length was 8.
      The maximum record length was 54.
```

Tip—Windows: To get a listing of just the names of the files and directories without all the extra information, use the dir /b windows command:

    filename os_cmd pipe "dir /b c:\temp";

Tip—Unix: To get a listing of just the names of the files and directories without all the extra information, use the ls -1 command (that is the number one, not a lowercase L after the hyphen):

    filename os_cmd pipe "ls -1 /tmp";

Working with the File System

From a DATA step, you might need to create external files or create a directory structure if it doesn't already exist. Rather than using system commands, you can use SAS DATA step functions to do this.

Most of the following examples were run on a Windows system and would need a little modification to work on a Unix system.

Creating and Deleting Directories

A set of DATA step functions is available for working on directories. You can check to see whether a directory exists (FEXIST() or FILEEXIST()), you can create a new directory (DCREATE()), you can delete a directory (FDELETE()), you can get the members of a directory (DNUM(), dread()) and you can look at information about a directory (DOPTNUM(), DOPTNAME(), DINFO()).

The example in Program 3.4 uses some of these functions to create a directory tree based on values from the sashelp.cars table. It creates a top level cars directory, and then a directory for each origin value. Under each origin, it creates a directory for each type of vehicle. SAS Log 3.5 shows the directories that are created.

Program 3.4: Creating Directories

```
proc sort data = sashelp.cars
     out = cars
     nodupkey;
   by origin type;
run;

data _null_;
   set cars;
   by origin type;

   length originDir topDir $200;
   retain originDir topDir;
```

```
   if (_n_ eq 1) then
   do;
      topDir = dcreate("cars", "c:\");
      if (topDir eq "") then
         topDir = "c:\cars";
      else
         putlog "Created: " topDir;
   end;

   if (first.origin) then
   do;
      originDir = "";
      originDir = dcreate(strip(origin), topDir);
      if (originDir eq "") then
        originDir = catx("\", topDir, origin);
      else
        putlog "Created: " originDir;
   end;

   typeDir = dcreate(strip(type), originDir);
   if (typeDir eq "") then
      typeDir = catx("\", originDir, type);
   else
      putlog "Created: " typeDir;

run;
```

SAS Log 3.5: Creating Directories

```
Created: c:\cars
Created: c:\cars\Asia
Created: c:\cars\Asia\Hybrid
Created: c:\cars\Asia\SUV
Created: c:\cars\Asia\Sedan
Created: c:\cars\Asia\Sports
Created: c:\cars\Asia\Truck
Created: c:\cars\Asia\Wagon
Created: c:\cars\Europe
Created: c:\cars\Europe\SUV
Created: c:\cars\Europe\Sedan
Created: c:\cars\Europe\Sports
Created: c:\cars\Europe\Wagon
Created: c:\cars\USA
Created: c:\cars\USA\SUV
Created: c:\cars\USA\Sedan
Created: c:\cars\USA\Sports
Created: c:\cars\USA\Truck
Created: c:\cars\USA\Wagon
```

Here is something to watch out for with DCREATE(): Note that the originDir is set to blank before the DCREATE(). function executes. This is because if it has a non-blank value and the DCREATE() function does not create the directory, then the originDir does not change. So you can't test for blank to see whether DCREATE() was successful.

If you want to delete a directory, you can use the FDELETE() function. However, the fdelete() can only be used to delete an empty directory, so you must first delete all the member directories and files. The DNUM() function (get the number of members) and the DREAD() function (get the member names) can be used to identify all the members to delete. As an example, the code in Program 3.5 deletes the cars directory structure that was created by Program 3.4. SAS Log 3.6 shows the directories as they are deleted.

Program 3.5: Deleting Directories

```
data _null_;
   carspath = "c:\cars";

   if (fileexist(carsPath));    ❶

   rc = filename("cars", carsPath);    ❷
   cars_did = dopen("cars");

   do c = 1 to dnum(cars_did);    ❸

      origin = dread(cars_did, c);    ❹
      originPath = catx("\", carsPath, origin);
      rc = filename("origin", originPath);
      origin_did = dopen("origin");

      do o = 1 to dnum(origin_did);    ❺
         type = dread(origin_did, o);    ❻
         typePath = catx("\", originPath, type);
         rc = filename("type", typePath);
         rc = fdelete("type");
         rc = filename("type");
         putlog "Deleted: " typePath;
      end; /* do o - loop through origin directories */

      origin_did = dclose(origin_did);    ❼
      rc = fdelete("origin");
      rc = filename("origin");
      putlog "Deleted: " originPath;

   end; /* do c - loop through cars directory */

   cars_did = dclose(cars_did);    ❽
   rc = fdelete("cars");
   rc = filename("cars");
   putlog "Deleted: " carsPath;

run;
```

❶ Process only if the cars directory exists.
❷ Open the cars directory.
❸ Loop through the members of the cars directory (the origin directories).
❹ Open the origin directory.
❺ Loop through the members of the origin directory (the type directories).
❻ Open the type directory and delete it.
❼ Delete the origin directory.
❽ Delete the cars directory.

SAS Log 3.6: Deleting Directories

```
Deleted:  c:\cars\Asia\Hybrid
Deleted:  c:\cars\Asia\Sedan
Deleted:  c:\cars\Asia\Sports
Deleted:  c:\cars\Asia\SUV
Deleted:  c:\cars\Asia\Truck
Deleted:  c:\cars\Asia\Wagon
Deleted:  c:\cars\Asia
Deleted:  c:\cars\Europe\Sedan
Deleted:  c:\cars\Europe\Sports
Deleted:  c:\cars\Europe\SUV
Deleted:  c:\cars\Europe\Wagon
Deleted:  c:\cars\Europe
Deleted:  c:\cars\USA\Sedan
Deleted:  c:\cars\USA\Sports
Deleted:  c:\cars\USA\SUV
Deleted:  c:\cars\USA\Truck
Deleted:  c:\cars\USA\Wagon
Deleted:  c:\cars\USA
Deleted:  c:\cars
```

See Also: The code in this example presumes that you know the exact depth of your directory structure. The Chapter 7 section headed "Simulating Recursion" contains an example showing how to traverse a directory structure when you don't know what is in the directory.

Working with Files

A set of functions is available for reading and writing external files (for example,FPUT(), FWRITE(), FGET(), FREAD()). I find using these functions to be a much more difficult way to read or write an external file than to use the FILE and INFILE statements with the PUT and INPUT statements. If you need to read or write an external file from outside a DATA step using a macro, then these file functions with %SYSFUNC work. From within a DATA step, I haven't found much use for these functions.

There are some other functions that I find much more useful for manipulating files. You can delete files (FDELETE()), copy files (FCOPY()), look at file information (FOPTNUM(), FOPTNAME(), FINFO()), and check to see if a file exists (FEXIST(), FILEEXIST()).

The example in Program 3.6 and SAS Log 3.7 uses these file functions to see whether a file exists, to copy one file to another, to create a file and write to it, to look at the file information, and finally to delete the file.

Program 3.6: Working with External Files

```
data _null_;

   oldFile = "c:\old.txt";
   newFile = "c:\new.txt";
   rc = filename("new", "newFile"); ❶

   if (fileexist(oldFile) and not fexist("new")) then ❷
   do;
      rc = filename("old", oldFile);
      rc = fcopy("old", "new"); ❸
      rc = filename("old");
      putlog "Copied old file (" oldFile ") to "
             "new file (" newFile +(-1) ")";
   end; /* if - old file exists, new file doesn't exist */

   else if (not fexist("new")) then ❹
   do;
      fid = fopen("new", "O"); ❺
      rc = fput(fid, "Testing");
      rc = fwrite(fid);
      rc = fclose(fid);
      putlog "Created new file (" newFile +(-1) ")";
   end; /* else if - old file and new file don't exist */

   fid = fopen("new");       ❻
   do i = 1 to foptnum(fid);
      name = foptname(fid, i);
      value = finfo(fid, name);
      putlog name= value=;
   end; /* do i - loop through file info */
   fid = fclose(fid);

   rc = fdelete("new");   ❼
   putlog "Deleted new file (" newFile +(-1) ")";

run;
```

❶ Use the FILENAME() function to set a fileref pointing to the new file (which does not exist).
❷ Use the FILEEXIST() function to see whether the old file exists, and use the FEXIST() function to see whether the new file exists. The difference between them is that FILEEXIST() accepts a path, while FEXIST() accepts a fileref. In this example, the old.txt file did not exist, so this IF statement was not true.
❸ Copy the old.txt file to the new.txt file.
❹ Check to see whether the new.txt file does not exist.
❺ Use FPUT() and FWRITE() to write to the new.txt file.
❻ Open the new.txt file and write all the available information about the file to the SAS log.
❼ Delete the new.txt file.

SAS Log 3.7: Working with External Files

```
Created new file (c:\new.txt)
name=Filename value=C:\Users\sasmih\newFile
name=RECFM value=V
name=LRECL value=32767
name=File Size (bytes) value=9
name=Last Modified value=04Feb2017:16:54:15
name=Create Time value=04Feb2017:16:54:15
Deleted new file (c:\new.txt)
```

Note: For simplicity in the file and directory function examples, I have left out the return code checking. To avoid errors and unexpected results, you should check for non-zero return codes from these functions. If the function fails, you can use the SYSMSG() function to capture the error message.

Reading and Writing External Files

If you are reading from and writing to an external file using the INFILE/INPUT and FILE/PUT statements, a couple of features can make things easier.

Using the $VARYING. Format and Informat

The $VARYING. format is a great way to write values that aren't always the same length. The syntax for this format requires that you specify the actual length of the value after $VARYING:

```
len = 3;
put var $varying. len;
```

The example in Program 3.7 shows the difference between using a regular $CHAR. format and $VARYING. format.

Program 3.7: Comparison of $CHAR. to $VARYING.

```
data _null_;
   set sashelp.class;
   file "c:\class.txt";
   put "$char   " name $char.;
   len = length(name);
   put "$varying " name $varying. len;
run;
```

Figure 3.3 shows a listing of the output file that is created. In the output file, I have selected all the text. You can see that when using the $CHAR. format, the line is padded with blanks, while the $VARYING. line is not. In this example, it doesn't make too much difference, but if you are creating a large file with long, varying lines of text, the extra spaces make the file even larger.

Figure 3.3: Comparison of $CHAR. to $VARYING.

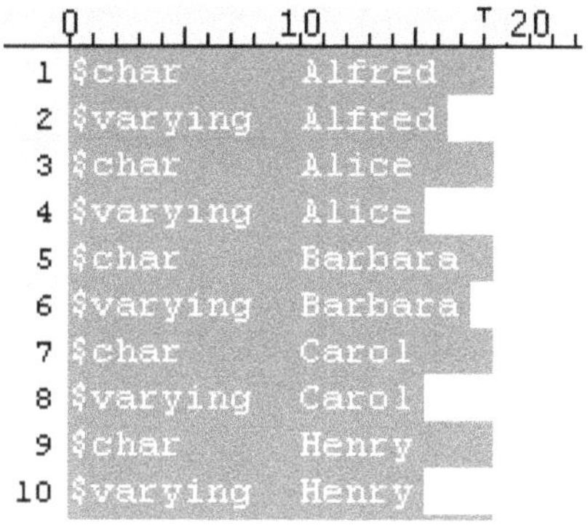

I use the $VARYING. format when I am generating code with a DATA step. It allows me to have some very long lines as well as very short lines, so that when viewed in another editor that has wrapping turned on, only the truly long lines wrap, not the short lines that are padded with blanks.

You can also use the $VARYING. informat when reading from a file. When you do this, you can use the RECLEN= option on the INFILE statement. This option creates an automatic variable that contains the length of the incoming line. You can then use that value to determine the length for the $VARYING. informat. Reading the data this way ensures that you don't read past the end of the input line. However, I think using the TRUNCOVER option on the INFILE statement is the easiest way to handle reading past the end of the line, so I rarely use the $VARYING. informat.

Using the FILEVAR= Option

The FILEVAR= option can be used on an INFILE statement or on a FILE statement. It lets you change the input file or output file during the execution of a DATA step without having multiple INFILE or FILE statements. It lets you write (using FILE) and read (using INFILE) multiple files in a single DATA step. So instead of creating a separate DATA step to read each file, and then appending the contents together, you can just use one DATA step.

The example in Program 3.8 and SAS Log 3.8 creates three files in a directory and writes to each of them in a single DATA step.

Program 3.8: Using FILEVAR= with FILE to Create Files

```
data _null_;

   dirName = "c:\filevar\";
   if (not fileexist(dirName)) then
      dirName = dcreate("filevar", "c:\");

   rc = filename("fvar", dirName);

   do f = 1 to 3;

      fname = cats(dirName, "file", f, ".txt");
      file fff filevar = fname;
      do i = 1 to f;
         put "Line " i +2 "file" f +(-1)".txt";
      end; /* do i - loop through records in file */

   end; /* do f - loop through files to be created */

   rc = filename("fvar");

run;
```

SAS Log 3.8: Using FILEVAR= with FILE to Create Files

```
NOTE: The file FFF is:
      Filename=c:\filevar\file1.txt,
      RECFM=V,LRECL=32767,File Size (bytes)=0,

NOTE: The file FFF is:
      Filename=c:\filevar\file2.txt,
      RECFM=V,LRECL=32767,File Size (bytes)=0,

NOTE: The file FFF is:
      Filename=c:\filevar\file3.txt,
      RECFM=V,LRECL=32767,File Size (bytes)=0,

NOTE: 1 record was written to the file FFF.
      The minimum record length was 18.
      The maximum record length was 18.
NOTE: 2 records were written to the file FFF.
      The minimum record length was 18.
      The maximum record length was 18.
NOTE: 3 records were written to the file FFF.
      The minimum record length was 18.
      The maximum record length was 18.
```

After Program 3.8 creates the files, Program 3.9 reads all the files. SAS Log 3.9 shows the results of reading the files.

Program 3.9: Using FILEVAR= with INFILE to Read Files

```
data _null_;

   if (fileexist("c:\filevar"));

   rc = filename("fvar", "c:\filevar");
   did = dopen("fvar");

   do f = 1 to dnum(did);

       fname = cats("c:\filevar\", dread(did, f));
       infile fff filevar = fname truncover end = eof;

       do until(eof);
           input @6 lineNum @9 filename $10.;
           put lineNum= filename=;
       end; /* do until - loop through the infile records */

   end; /* do f - loop through files in directory */

   did = dclose(did);
   rc = filename("fvar");
   stop;

run;
```

SAS Log 3.9: Using FILEVAR= with INFILE to Read Files

```
NOTE: The infile FFF is:
      Filename=c:\filevar\file1.txt,
      RECFM=V,LRECL=32767,File Size (bytes)=20,

lineNum=1 filename=file1.txt
NOTE: The infile FFF is:
      Filename=c:\filevar\file2.txt,
      RECFM=V,LRECL=32767,File Size (bytes)=40,

lineNum=1 filename=file2.txt
lineNum=2 filename=file2.txt
NOTE: The infile FFF is:
      Filename=c:\filevar\file3.txt,
      RECFM=V,LRECL=32767,File Size (bytes)=60,

lineNum=1 filename=file3.txt
lineNum=2 filename=file3.txt
lineNum=3 filename=file3.txt
NOTE: 1 record was read from the infile FFF.
      The minimum record length was 18.
      The maximum record length was 18.
NOTE: 2 records were read from the infile FFF.
      The minimum record length was 18.
      The maximum record length was 18.
NOTE: 3 records were read from the infile FFF.
      The minimum record length was 18.
      The maximum record length was 18.
```

> **Warning**: Do not forget the STOP statement at the end. If you leave it off, the DATA step goes into an infinite loop that will drive you crazy trying to debug. If you get into an infinite loop, always remember to check that you have the STOP statement.

Reading Date and Time Values

The ANYDTDTE. (for dates), ANYDTDTM. (for datetimes), and ANYDTTME. (for times) informats are incredibly useful when reading in data that has date and time values. You can use these informats to read many date and time formats, instead of having to find the informat that corresponds to your data. In addition, if you have a date column that has values in different formats (for example, mm/dd/yyyy and ddMONyy), or you don't know what format the data will be in, you can read them all with the ANYDTDTE. informat.

Program 3.10 and SAS Log 3.10 show how to use the ANYDT* informats.

Program 3.10: Using ANYDT* Informats to Read Date and Time Values

```
data _null_;
   format date date9.
          time time5.
          datetime datetime18.;
   input @1 date anydtdte15.
         @17 time anydttme13.
         @31 datetime anydtdtm21.;
   putlog date= time= datetime=;
datalines;
01JAN17      2:00pm     01Jan17:14:00
1/1/2017     14:00:00    170101 02:00 pm
January 1, 2017 14:00        1/1/2017 14:00
17001        02:00:00pm  01JAN2017
01Jan17:14:00  01Jan17:14:00 01Jan17:14:00
;
```

SAS Log 3.10: Using ANYDT* Informats to Read Date and Time Values

```
date=01JAN2017 time=14:00 datetime=01JAN17:14:00:00
date=01JAN2017 time=14:00 datetime=01JAN17:14:00:00
date=01JAN2017 time=14:00 datetime=01JAN17:14:00:00
date=01JAN2017 time=14:00 datetime=01JAN17:00:00:00
date=01JAN2017 time=14:00 datetime=01JAN17:14:00:00
```

Creating a CSV File from a Data Table

There are several ways of creating a comma-separated values (CSV) file from a data table, but one of the simplest might be one that you haven't heard of: %DS2CSV. This macro utility is part of Base SAS, and all you need to do is tell it the name of the data table and the name of the CSV file, and run it. There are lots of documented options, including options that can be used if you are running from a web application and want to allow the user to download the CSV file. Program 3.11 and SAS Log 3.11 show a simple usage of %DS2CSV to create a CSV file from the sashelp.class table.

Program 3.11: Using %DS2CSV to Create a CSV File

```
%ds2csv(data = sashelp.class,
        csvfile = c:\temp\class.csv,
        runmode = b);
```

SAS Log 3.11: Using %DS2CSV to Create a CSV File

```
NOTE: CSV file successfully generated for SASHELP.CLASS.
```

You need to set the RUNMODE= option to b if you are running the code in batch or interactive SAS; otherwise, the macro will attempt to run in a SAS workspace server.

PROC EXPORT is another easy way to create a CSV file. It has plenty of documented options that will enable you to create the CSV the way you want it. It also generates a SAS DATA step that you can save and modify if the code isn't quite what you want. Program 3.12 and SAS Log 3.12 show how to use PROC EXPORT to create a CSV file from the sashelp.class table.

Program 3.12: Using PROC EXPORT to Create a CSV File

```
proc export data = sashelp.class
            outfile = "c:\temp\class.csv"
            dbms = csv
            label
            replace;
run;
```

SAS Log 3.12: Using PROC EXPORT to Create a CSV File

```
7    /*****************************************************************
8    *    PRODUCT:   SAS
9    *    VERSION:   9.4
10   *    CREATOR:  External File Interface
11   *    DATE:     05MAY17
12   *    DESC:     Generated SAS Datastep Code
13   *    TEMPLATE SOURCE: (None Specified.)
14   *****************************************************************/
15        data _null_;
16        %let _EFIERR_ = 0; /* set the ERROR detection variable */
17        %let _EFIREC_ = 0; /* clear export record count variable */
18        file 'c:\temp\class.csv' delimiter=',' DSD DROPOVER
lrecl=32767;
19        if _n_ = 1 then  /* write column names or labels */
20         do;
21           put
22              '"' "Name" '"'
23            ','
24              '"' "Sex" '"'
25            ','
26              '"' "Age" '"'
27            ','
28              '"' "Height" '"'
29            ','
30              '"' "Weight" '"'
31            ;
32         end;
33       set SASHELP.CLASS   end=EFIEOD;
```

```
34          format Name $8. ;
35          format Sex $1. ;
36          format Age best12. ;
37          format Height best12. ;
38          format Weight best12. ;
39      do;
40          EFIOUT + 1;
41          put Name $ @;
42          put Sex $ @;
43          put Age @;
44          put Height @;
45          put Weight ;
46          ;
47      end;
48      if _ERROR_ then call symputx('_EFIERR_',1); /* set ERROR
detection macro variable */
49      if EFIEOD then call symputx('_EFIREC_',EFIOUT);
50      run;

NOTE: The file 'c:\temp\class.csv' is:
      Filename=c:\temp\class.csv,
      RECFM=V,LRECL=32767,File Size (bytes)=0,
      Last Modified=06May2017:08:35:33,
      Create Time=17Apr2017:11:31:58

NOTE: 20 records were written to the file 'c:\temp\class.csv'.
      The minimum record length was 17.
      The maximum record length was 36.
NOTE: There were 19 observations read from the data set
SASHELP.CLASS.
NOTE: DATA statement used (Total process time):
      real time        0.02 seconds
      cpu time         0.03 seconds


19 records created in c:\temp\class.csv from SASHELP.CLASS.

NOTE: "c:\temp\class.csv" file was successfully created.
NOTE: PROCEDURE EXPORT used (Total process time):
      real time        0.16 seconds
      cpu time         0.09 seconds
```

Reading a CSV File with Embedded Line Feeds

A common issue when you are working in a Microsoft Excel spreadsheet is using Alt+Enter to go to a new line within the same cell thus creating an embedded line feed character. If you then turn this Excel spreadsheet into a CSV, the value contains the embedded line feed. If you view the file in some basic text editors, you will see that it goes to a new line at this point. If you try to read this file with SAS software, it will see the line feed character as an end of record, will set the rest of the fields to blank, and will go to the next iteration of the DATA step and start reading from the middle of the line.

If you are using a DATA step with an INFILE statement and INPUT statements, then this is an easy fix. You can add the TERMSTR= option to the INFILE statement and it will handle the line feed character correctly.

For this example, I used the code in Program 3.13 to create a CSV file that has an embedded line feed character (0Ax).

Program 3.13: Creating a CSV with an Embedded Line Feed

```
data _null_;
    file "c:\temp\test.csv";
    put '"aaa","bbb","ccc","ddd"';
    put '"aa","bb","c' '0A'x 'c","dd"';
    put '"a","b","c","d"';
run;
```

If you look at the file in an editor, you will see that the second record is split over two lines. Figure 3.4 shows the file in Notepad.

Figure 3.4: Creating a CSV with an Embedded Line Feed

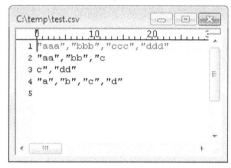

Program 3.14 and Figure 3.5 show what happens when you attempt to read this file with a DATA step and INFILE statement.

Program 3.14: Attempting to Read a CSV File with Standard INFILE/INPUT

```
data test;
    infile "c:\temp\test.csv" dlm = "," dsd truncover;
    length a b c d $5;
    input a b c d;
run;
```

Figure 3.5: Attempting to Read a CSV File with Standard INFILE/INPUT

Obs	a	b	c	d
1	aaa	bbb	ccc	ddd
2	aa	bb	"c	
3	c"	dd		
4	a	b	c	d

This table is definitely not what you wanted. So, you can add the TERMSTR=CRLF option to the INFILE statement. This tells the INFILE statement to only use a carriage return + line feed as the end of record. Program 3.15 and Figure 3.6 show what happens when you use the TERMSTR=CRLF option.

Program 3.15: Reading a CSV File with the TERMSTR= Option

```
data test;
    infile "c:\temp\test.csv" dlm = "," dsd truncover
                                termstr = CRLF;
    length a b c d $5;
    input a b c d;
run;
```

Figure 3.6: Reading a CSV File with the TERMSTR= Option

Obs	a	b	c	d
1	aaa	bbb	ccc	ddd
2	aa	bb	c c	dd
3	a	b	c	d

You can see that the output table is now correct.

However, if you want to read the CSV file with PROC IMPORT, or you actually have an embedded carriage return (0Dx) + line feed (0Ax), not just a line feed character in the file, the only way to handle it is to preprocess the file and change the line feed characters to another character. Additionally, I usually want to maintain the line feeds in my data, so I convert each line feed character to a value that I will be able to find again. After the data has been read, I can turn those values back into line feeds. If you don't care about maintaining the line feed, then you can just convert them to blanks.

Program 3.16 shows the code that preprocesses the file and converts any line feed that is part of a string that is enclosed in double quotation marks into "~~". Figure 3.7 shows the preprocessed file in Notepad.

Program 3.16: Preprocessing a CSV File and Replacing Line Feeds

```
data _null_;
    infile "c:\temp\test.csv" recfm=n sharebuffers;
    file "c:\temp\test_TEMP.csv" recfm=n;
    input a $char1.;
    retain inQuotes 0;
    if a = '"' then
        inQuotes = (not inQuotes);
    if (a eq '0A'x and inQuotes) then
        put '~~';
    else
        put a $char.;
run;
```

Figure 3.7: Preprocessing a CSV File and Replacing Line Feeds—test_TEMP.csv

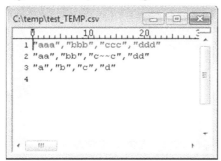

Now Program 3.17 can read the test_TEMP.csv file into a SAS table and put the line feeds back where they should be. Figure 3.8 shows the data table that is created from the temporary CSV file.

Program 3.17: Reading a Preprocessed CSV File

```
data test;
    infile "c:\temp\test_TEMP.csv" dlm = "," dsd truncover;
    length a b c_in c c_hex d $10;
    input a b c_in d;
    c = tranwrd(c_in, "~~", "0A"x);
    c_hex = put(c, hex6.);
run;
```

Figure 3.8: Reading a Preprocessed CSV File

Obs	a	b	c_in	c	c_hex	d
1	aaa	bbb	ccc	ccc	636363	ddd
2	aa	bb	c~~c	c c	630A63	dd
3	a	b	c	c	632020	d

As you can see from the output table, the second row shows that c_in (the value from the input file) contains "~~", and then c has the value with the "~~" replaced by a line feed. The c_hex column shows the hexadecimal representation of c, and you can see the 0A in the middle. After the file has been preprocessed, you can read it with whatever tool you want and make any updates that you need.

Chapter 4: The Macro Facility

Introduction

I rarely write a SAS program that doesn't contain macro variables and macros. Macros enable you to take more control over your programs: passing values from one part of the program to another, conditionally executing sections of code, running the same piece of code multiple times, and performing countless other tasks. Here are some tips for using macros to their full potential. If you are not a macro user, maybe some of this will help you embrace them.

Understanding Macro Variables

Macro variables are great for moving single pieces of information around your program—in and out of DATA steps, procedures, and macros. I can't imagine writing a SAS program without a single macro variable.

Using the "Dot" in Macro Variable Names

When do you use a dot at the end of a macro variable name? Having seen lots of SAS code, I find programmers of all levels using the dot somewhat randomly with their macro variables. If you are using a dot at the end of most of your macro variables, feel free to keep doing so; it won't do any harm. However, here is an explanation of when you should use a dot and when you shouldn't.

You need to put a dot at the end of the macro variable name if the character directly following the macro variable name could be part of the name. So if you have any letter, number, or underscore directly after the macro variable name, then you need to add the dot so that the SAS program knows where the macro variable name ends. The only exception is when you have a dot directly after the macro variable name. In that case you need to add a second dot as the end-of-name marker.

Program 4.1 and SAS Log 4.1 show some examples of when to use the dot.

Program 4.1: Examples of Using Dots

```
%let a = 123;
%let b = xyz;
%let value = &a&b;   /* dots are not required */ ❶
%put &=value;
%let value = &a.abc; /* dot is required */ ❷
%put &=value;
%let value = &a abc; /* dot is not required */ ❸
%put &=value;
%let value = &a(abc); /* dot is not required */ ❹
%put &=value;
%let value = &a..abc; /* double dots to keep one dot */ ❺
%put &=value;
```

❶ A dot is not required after &a because the next character, "&", cannot be part of the macro variable name. A dot is not required after &b because the next character, ";", cannot be part of the macro variable name.

❷ A dot is required after &a because the next character, "a", could be part of the macro variable name.

❸ A dot is not required after &a because the next character, " " (blank), cannot be part of the macro variable name.

❹ A dot is not required after &a because the next character, "(", cannot be part of the macro variable name.

❺ Two dots are required in order to keep a dot between the macro variable and the "abc" string.

SAS Log 4.1: Examples of Using Dots

```
1 %let a = 123;
2 %let b = xyz;
3 %let value = &a&b;   /* dots are not required */
4 %put &=value;
VALUE=123xyz
5 %let value = &a.abc;  /* dot is required */
6 %put &=value;
VALUE=123abc
7 %let value = &a abc;  /* dot is not required */
8 %put &=value;
VALUE=123 abc
9 %let value = &a(abc); /* dot is not required */
10 %put &=value;
VALUE=123(abc)
11 %let value = &a..abc; /* double dots to keep one dot */
12 %put &=value;
VALUE=123.abc
```

Determining How Many &s to Use

Have you ever written or seen code with a string of &s in front a macro variable name? How many do you need and why?

The simplest use is when you are simulating an array with macro variables: &var1, &var2, &var3, and &var4. If you want to loop through these values, you can do something like this:

```
%do i = 1 %to 4;
    %let x = &&var&i;
%end;
```

If you just use &var&i, the SAS compiler sees this as two separate macro variables: &var and &i. To get the macro array value, you need the SAS compiler to first resolve the &i to its value (a number), and then resolve &var#. So you use two &s in front of var and a single & in front of the i to hang on to var until i has been resolved.

The SAS software reads each statement from left to right. Each time it sees two &s, it converts them to a single &, and each time it sees a single &, it converts the macro variable to its value. Then it reads through the statement again and converts two &s to a single & and resolves any single &s into the value. It continues in this way until all the macro variables have been resolved.

Here are several examples to illustrate this process.

Example 1:

```
%let i = 1;
%let var1 = abc;
%let x = &&var&i;
```

The SAS software resolves the last statement starting at the left:

1. Resolve the && to a single & (&var), and then resolve &i to 1. So the statement becomes
   %let x = &var1;
2. Resolve &var1 to abc to get
   %let x = abc;
3. No more macro variables are left, so the process is finished.

Example 2:

```
%let i = 1;
%let nm = var;
%let var1 = abc;
%let x = &&&nm&i;
```

The SAS software resolves the last statement starting at the left:

1. Resolve the first && to a single &, then resolve &nm to var, and then resolve &i to 1. So the statement becomes

   %let x = &var1;

2. Resolve &var1 to abc to get

   %let x = abc;

3. No more macro variables are left, so the process is finished.

Example 3:

```
%let i = 1;
%let a1 = 2;
%let var2 = abc;
%let x = &&&&var&&a&i;
```

The SAS software resolves the last statement starting at the left:

1. Resolve the first && to a single &, then the second && to a single & (&&var), then the third && to a single & (&a), and then &i to 1. So the statement becomes

   %let x = &&var&a1;

2. Resolve the first && to a single &, (&var), and then resolve &a1 to 2. So the statement becomes

   %let x = &var2;

3. Resolve &var2 to abc to get

   %let x = abc;

4. No more macro variables are left, so the process is finished.

Understanding Quotation Marks

Quotation marks are not needed to define character strings in macros and macro variables. In fact, a single or double quotation mark is just another character in the macro world. (Well, that's not totally true: We are still working with the SAS compiler, and if you have a mismatched quotation mark, it causes serious problems in your program.)

Everything in macro is a character string, which is why it isn't necessary to tell the macro compiler that a value is a string by putting quotation marks around the value. This means that numbers are also strings, so if you want to do math using macro variables, you need to use the %EVAL() function.

As an example, this code is incorrect (unless the macro variable's value actually has quotation marks in it):

```
%if (&macroVar eq "TEST") %then
```

This code looks for a value that contains double quotation marks.

If you like quotation marks to set off your string values, you must use them on both sides of the comparison, because they are just other characters that are being compared. So, you can do this:

```
%if ("&macroVar" eq "TEST") %then
```

I prefer to just trust the macro compiler and not to use quotation marks:

```
%if (&macroVar eq TEST) %then
```

If you do need to have an unmatched quotation mark (single or double) in a value, you can use %STR() or %NRSTR() functions with the percentage sign (%) to escape the quotation mark:

```
%let macroVar = %str(TE%'ST);
```

You can use the % to escape unmatched parentheses as well.

Using Macro Quoting

I find that macro quoting is one of the most difficult parts of the macro facility for programmers to understand (including me). Because all macro values are strings that don't need to be quoted with quotation marks, you run into conflicts when the macro compiler tries to execute your string that you just want to be treated as a string of characters. The macro quoting functions are needed to tell the compiler to leave your string alone. There are lots of macro quoting functions, as well as plenty of documentation on how to use them. The functions that I use most are %STR() and %BQUOTE(), and as a last resort, I like %SUPERQ().

The %STR() Function

This is probably the most common quoting function, and I use it mostly on hardcoded strings. For example, suppose that you want to use a value in the %SCAN() function that contains a comma like this:

```
%let macroVar = %scan(test,test, 1, ,);
```

The macro compiler interprets the extra commas as parameter delimiters instead of strings and produces an error. This is a good time to use %STR() to hide the commas from the function:

```
%let macroVar = %scan(%str(test,test), 1, %str(,));
```

The %BQUOTE() Function

I most often use the %BQUOTE() function to quote macro variables that could contain a value that might be misinterpreted by the macro compiler. I often use this code to check for blank values in parameters that are coming into a macro:

```
%if (%bquote(&macroVar) eq ) %then
```

I have found that %BQUOTE() will handle the majority of issues with strange values.

The %SUPERQ() Function

One other quoting function that I like to use as a last resort is the %SUPERQ() function. This function masks everything, including embedded &s and %s. Its syntax is different from that of other functions:

```
%if (%superq(macroVar) eq ) %then
```

That is not a typographical error. The macro variable in the %SUPERQ() function should not have an & at the beginning. The only parameter that you can pass to %SUPERQ() is a macro variable name. You cannot enclose other text in the function for quoting as you can with the other quoting functions. The %SUPERQ() function is useful if you want to mask unbalanced quotation marks, unbalanced parentheses, and &s and %s. If the value that you are testing could have any of these characters, %SUPERQ() might be the best way to mask everything.

Unquoting

When you use a macro quoting function on a value, there are hidden macro quotation marks that are added to the value. The macro facility can handle these hidden quotation marks, but regular SAS code doesn't always compile correctly. You can end up with a situation like the one in Program 4.2 and SAS Log 4.2:

Program 4.2: Quoted Macro Value Causes an Error

```
%let string = %str(%"I%'m testing quotes%");
data _null_;
   x = &string;
   putlog x=;
run;
```

SAS Log 4.2: Quoted Macro Value Causes an Error

```
1    %let string = %str(%"I%'m testing quotes%");
2    data _null_;
3       x = &string;
NOTE: Line generated by the macro variable "STRING".
1       "I'm testing quotes"
        -
        386
        -
        200
        76
ERROR 386-185: Expecting an arithmetic expression.

ERROR 200-322: The symbol is not recognized and will be ignored.

ERROR 76-322: Syntax error, statement will be ignored.

4       putlog x=;
5    run;
```

When this happens, you need to unquote the macro variable when you use it, as shown in Program 4.3 and SAS Log 4.3:

Program 4.3: Unquoting a Macro Value

```
%let string = %str(%"I%'m testing quotes%");
data _null_;
   x = %unquote(&string);
   putlog x=;
run;
```

SAS Log 4.3: Unquoting a Macro Value

```
1    %let string = %str(%"I%'m testing quotes%");
2    data _null_;
3        x = %unquote(&string);
4        putlog x=;
5    run;

x=I'm testing quotes
```

So if you are using macro variables in regular SAS code and you get syntax errors that don't make sense—everything *looks* right—try unquoting the macro variable.

Forcing Single Quotation Marks

If you want to resolve a macro variable's value inside quotation marks, you must use double quotation marks. Sometimes (for example, when using pass-through in PROC SQL—you need the value to be in single quotation marks. Because it won't resolve inside of the single quotation marks, you need to embed the single quotation marks into the value. Program 4.4 and SAS Log 4.4 show one way of doing this.

Program 4.4: Creating a Macro Value Surrounded by Single Quotation Marks

```
%let value = ABC;
%let quotedValue = %str(%')&value%str(%');
%put &=quotedValue;
```

SAS Log 4.4: Creating a Macro Value Surrounded by Single Quotation Marks

```
QUOTEDVALUE='ABC'
```

If you use this syntax, you must use the %UNQUOTE() function when you use &quotedValue in non-macro code to remove the macro quoting that is added by the %STR() function. Program 4.5 and SAS Log 4.5 show an example of using the value without unquoting it (which causes errors), and an example with unquoting (which works correctly).

Program 4.5: Using the Single Quoted Macro Value

```
%let value = ABC;
%let quotedValue = %str(%')&value%str(%');
%put &=quotedValue;

data _null_;
   x = &quotedValue;
   putlog x=;
run;

data _null_;
   x = %unquote(&quotedValue);
   putlog x=;
run;
```

SAS Log 4.5: Using the Single Quoted Macro Value

```
QUOTEDVALUE='ABC'
4
5    data _null_;
6        x = &quotedValue;
NOTE: Line generated by the macro variable "QUOTEDVALUE".
1      'ABC'
       -
       386
       -
       200
ERROR 386-185: Expecting an arithmetic expression.

ERROR 200-322: The symbol is not recognized and will be ignored.

7        putlog x=;
8    run;

NOTE: The SAS System stopped processing this step because of errors.
NOTE: DATA statement used (Total process time):
      real time          0.00 seconds
      cpu time           0.01 seconds


9
10   data _null_;
11       x = %unquote(&quotedValue);
12       putlog x=;
13   run;

x=ABC
```

A simpler way to get single quotation marks around the value is to use the %TSLIT macro function. When you use this function, it adds single quotation marks to the value, and you won't need to unquote it. Program 4.6 and SAS Log 4.6 show how to use %TSLIT.

Program 4.6: Using the %TSLIT() Function to Add Single Quotation Marks

```
%let value = ABC;
%let quotedValue = %tslit(&value);
%put &=quotedValue;

data _null_;
   x = &quotedValue;
   putlog x=;
run;
```

SAS Log 4.6: Using the %TSLIT() Function to Add Single Quotation Marks

```
QUOTEDVALUE='ABC'
4
5    data _null_;
6        x = &quotedValue;
7        putlog x=;
8    run;

x=ABC
```

Writing Macros

Macro programming is different from most programming languages that I've worked with, but it is worth getting to know because it makes your SAS programs much more effective. Read *Carpenter's Complete Guide to the SAS Macro Language* by Art Carpenter to become an expert. In the meantime, here are some tips to help improve your programs.

Using Autocall Macros

Autocall macros are defined in files on the file system or in a SAS catalog. After you set the SASAUTOS option to point to your autocall macro library, you can call the macros from a SAS program without having to include and submit the code first. In order to use the autocall facility, you must turn on the MAUTOSOURCE option:

```
options mautosource;
```

I prefer saving my macros in files rather than compiling them into a SAS catalog, because they are easier to maintain. But catalogs are very useful if you want to compile the macro without the source code so that your macros are protected from modification and plagiarism.

Storing Autocall Macros in Files

If you have stored your macros in .sas files, you tell the SAS software about them by issuing a filename statement that points to the directory where the files are located and an options statement that adds your library to the SASAUTOS option:

```
filename mymacros "c:\MyMacros";
options insert = (sasautos = mymacros);
```

The INSERT= option adds your fileref to the beginning of the list of SASAUTOS libraries, which means that your directory is searched first. You can also use the APPEND= option to add your fileref to the end of the list of SASAUTOS libraries. After the SASAUTOS option has been set, you can execute any of the macros that are stored in your directory.

Here are a few guidelines for using files:

- Include only a single macro in each file. The %MACRO statement should be at the beginning of the file and the %MEND statement should be at the end of the file. The autocall facility enables you to have multiple macros in a file, but then the maintenance gets complicated, because the macros are difficult to locate. It is inefficient to nest macros within other macros, because the nested macro is recompiled every time the primary macro is executed. Best practice is to put every macro in its own autocall file. I use the parent macro's name in the sub-macros' names to indicate that a macro belongs to another macro.

- The filename must match the name of the macro. So if your macro name is delete_tables, then the file must be named delete_tables.sas. On Windows, the case of the autocall macro filename doesn't matter, and neither does the case of the name of the macro in the file, but on Unix, the filename must be lowercase. My suggestion is to always give your macros lowercase names and use that same case throughout your programs. It's easier to read, and you never have to think about it. It also makes moving code from Windows to Unix much easier. If you are using all lowercase macro names, then I also suggest using underscores to separate the words. Rather than a macro called %deletetables, I would call it %delete_tables.

- Be careful about using a macro name that has already been defined in SAS. There are many Base SAS autocall macros that generally have short names (8 or fewer characters), as well as other SAS products that have their own autocall macros. The product-specific macros will usually have a special prefix on the macro name (for example, SAS Data Integration Studio prefixes macros with etls_, and SAS Enterprise Guide uses eg_ as a prefix. You might want to have a project-specific prefix for your autocall macros that will keep you from clashing with SAS product macros.

- The SASAUTOS option is used to specify the location of your autocall macros. It can be specified at any time during your SAS program, or you can put it in the config file or in the statement that you use to invoke SAS. The SASAUTOS option can point to a single library, or it can point to multiple libraries. To avoid losing any of the product-defined autocall libraries from the SASAUTOS option, add your library to the beginning or end of the option value. The INSERT option adds the library to the beginning (so that your library will be searched first) and the APPEND option adds the library to the end (so that your library will be searched last).

Storing Autocall Macros in Catalogs

Generally, if you want to store your macros in a SAS catalog, you store them as compiled code. You need to specify the MSTORED system option to turn on the stored compiled macro facility:

```
options mstored;
```

You also need to specify the SASMSTORE= option to point to the SAS library where you want to save the compiled macros:

```
libname mymacros "c:\macroLibrary";
options sasmstore = mymacros;
```

Then add the STORE option to your %MACRO statement:

```
%macro delete_tables / store;
```

Finally, submit the macro to store the compiled macro in mymacros.sasmacro.delete_tables.macro. If you attempt to open this file, you will find that you can't view or edit it. If you want to keep the source code along with the compiled code, then you can add the SOURCE option to the macro statement:

```
%macro delete_tables / store source;
```

Then when you submit the macro, the source code is saved with the compiled code in the mymacros.sasmacro.delete_tables.macro catalog entry. To view the source, you can use the %COPY macro:

```
%copy delete_tables / source;
```

The %COPY macro writes the macro code to the SAS log. There is also an OUTFILE= option on %COPY that enables you to save the source code to an external file:

```
%copy delete_tables /
       source
       outfile = "c:\macroLibrary\delete_tables.sas";
```

This code creates a delete_tables.sas file with the macro code in it.

> **Warning:** If you don't use the SOURCE option when you store a compiled macro, it is not possible to get the source code from the compiled macro. So be sure to keep a copy of the source code somewhere.

To call a compiled macro, you need to have the MSTORED and SASMSTORE= options set:

```
libname mymacros "c:\macroLibrary";
options sasmstore = mymacros mstored;
```

Then you can call any of the macros that are stored in this catalog.

Determining When to Write a Macro

I have seen lots of different styles and philosophies on writing macros, and whatever makes sense to you is the right way. But if you haven't thought about it, think about it now.

I've seen some programmers who write lots of little macros throughout a SAS program. Whenever they need to do some macro code (for example, conditionally executing a piece of code), they create a macro for that section and then call the macro when they need it (often directly after the macro definition). So they might have a SAS program that looks like this:

```
%let endNum = 2;

data _null_;

   %macro count;
      %do i = 1 %to &endNum;
         putlog "line &i";
      %end;
   %mend;
   %count;

   putlog "After Count";

   %macro test;
      %if (&endNum ge 10) %then
      %do;
         putlog "Greater than or equal to 10";
```

```
        %end;
        %else
        %do;
            putlog "Less than 10";
        %end;
    %mend;
    %test;

run;
```

This program works correctly, but it is difficult to read with the multiple %MACRO and %MEND statements and the calls to the macros. I suggest that if you have some macro work to do in a program, then put all your code into one big macro. Then you can use macro statements whenever you need to. As a bonus, it is easy to move to an autocall library if you decide to do that.

Program 4.7 and SAS Log 4.7 show how I would write the previous code.

Program 4.7: One Big Macro Instead of Lots of Little Macros

```
%macro count_test;

    %let endNum = 2;

    data _null_;

        %do i = 1 %to &endNum;
            putlog "line &i";
        %end;

        putlog "After Count";

        %if (&endNum ge 10) %then
        %do;
            putlog "Greater than or equal to 10";
        %end;
        %else
        %do;
            putlog "Less than 10";
        %end;
    run;

%mend count_test;
%count_test;
```

SAS Log 4.7: One Big Macro Instead of Lots of Little Macros

```
line 1
line 2
After Count
Less than 10
```

This code does the same thing as the previous code, but it is easier to read and to write.

Another important use of macros is for code that needs to be run multiple times. If you have a section of code that you need to run for different scenarios, whether it is a little bit of DATA step code, or a whole series of procedures and DATA steps, then write a macro instead of duplicating the code. You can use parameters to provide the information each scenario needs. This makes maintenance much simpler, because you only need to modify a single section of code rather than multiple sections.

Program 4.8 and SAS Log 4.8 provide an example of a macro for a duplicate piece of DATA step code.

Program 4.8: Duplicate Piece of DATA Step Code in a Macro

```
%macro convert_to_centimeters(var);
    putlog &var= "in";
    &var = &var * 2.54;
    putlog &var= "cm";
%mend;

data _null_;
    set sashelp.fish (obs=2);
    putlog species=;
    %convert_to_centimeters(length1);
    %convert_to_centimeters(length2);
    %convert_to_centimeters(length3);
run;
```

SAS Log 4.8: Duplicate Piece of DATA Step Code in a Macro

```
Species=Bream
Length1=23.2 in
Length1=58.928 cm
Length2=25.4 in
Length2=64.516 cm
Length3=30 in
Length3=76.2 cm
Species=Bream
Length1=24 in
Length1=60.96 cm
Length2=26.3 in
Length2=66.802 cm
Length3=31.2 in
Length3=79.248 cm
```

Program 4.9 is an example of a macro that contains multiple procedures and is called for 3 different scenarios. Figure 4.1 shows the output table that is created.

Program 4.9: Complex Macro for Multiple Scenarios

```
%macro makeBucket(in_table=,
                  out_table=,
                  var=,
                  lower_value=,
                  upper_value=,
                  bucket_name=);

   proc sql noprint;

      create table bucket as
         select *,
                "&bucket_name" as bucket length = 20
            from &in_table
               %if (&lower_value eq ) %then
               %do;
                  where &var le &upper_value;
               %end;
               %else %if (&upper_value eq ) %then
               %do;
                  where &var gt &lower_value;
               %end;
               %else
               %do;
                  where &var gt &lower_value and
                        &var le &upper_value;
               %end;

      select sum(age) / count(*) into :avgAge
         from bucket;

      create table bucket2 as
         select *,
                &avgAge as average_age
            from bucket;

   quit;

   proc append base = &out_table
               data = bucket2;
   run;

   proc datasets lib = work nolist nowarn;
      delete bucket bucket2;
   quit;

%mend makeBucket;

/* get rid of the final table so we don't start with any data */
proc datasets lib = work nolist nowarn;
   delete class;
quit;
```

```
%makeBucket(in_table = sashelp.class,
            out_table = work.class,
            var = age,
            upper_value = 11,
            bucket_name = %str(<= 11));
%makeBucket(in_table = sashelp.class,
            out_table = work.class,
            var = age,
            lower_value = 11,
            upper_value = 13,
            bucket_name = %str(11-13));
%makeBucket(in_table = sashelp.class,
            out_table = work.class,
            var = age,
            lower_value = 13,
            bucket_name = %str(> 13));
```

Figure 4.1: Complex Macro for Multiple Scenarios

Obs	Name	Sex	Age	Height	Weight	bucket	average_age
1	Joyce	F	11	51.3	50.5	<= 11	11.0000
2	Thomas	M	11	57.5	85.0	<= 11	11.0000
3	Alice	F	13	56.5	84.0	11-13	12.3750
4	Barbara	F	13	65.3	98.0	11-13	12.3750
5	James	M	12	57.3	83.0	11-13	12.3750
6	Jane	F	12	59.8	84.5	11-13	12.3750
7	Jeffrey	M	13	62.5	84.0	11-13	12.3750
8	John	M	12	59.0	99.5	11-13	12.3750
9	Louise	F	12	56.3	77.0	11-13	12.3750
10	Robert	M	12	64.8	128.0	11-13	12.3750
11	Alfred	M	14	69.0	112.5	> 13	14.6667
12	Carol	F	14	62.8	102.5	> 13	14.6667
13	Henry	M	14	63.5	102.5	> 13	14.6667
14	Janet	F	15	62.5	112.5	> 13	14.6667
15	Judy	F	14	64.3	90.0	> 13	14.6667
16	Mary	F	15	66.5	112.0	> 13	14.6667
17	Philip	M	16	72.0	150.0	> 13	14.6667
18	Ronald	M	15	67.0	133.0	> 13	14.6667
19	William	M	15	66.5	112.0	> 13	14.6667

Wrapping Long Lines of Code

Sometimes you have a very long line of macro code, but wrapping to the next line might add extra spaces in a macro variable that you don't want. Program 4.10 and SAS Log 4.9 show an example of what happens when you wrap a line of code in the middle of a value.

Program 4.10: Wrapped Message String Causes Extra Blanks

```
%let msg = This is a
           message;
%put &=msg;
```

SAS Log 4.9: Wrapped Message String Causes Extra Blanks

```
MSG=This is a                message
```

There are several ways to make this work correctly. Program 4.11 and SAS Log 4.10 show how to fix it with multiple %LET statements.

Program 4.11: Multiple %let Statements

```
%let msg = This is a;
%let msg = &msg message;
%put &=msg;
```

SAS Log 4.10: Multiple %let Statements

```
MSG=This is a message
```

Program 4.12 and SAS log 4.11 show how to fix this with the %CMPRES() function.

Program 4.12: Using the %CMPRES() Function

```
%let msg = This is a %cmpres(
           ) message;
%put &=msg;
```

SAS Log 4.11: Using the %CMPRES() Function

```
MSG=This is a message
```

The %CMPRES() function gets rid of any blanks in the value enclosed by the parentheses, so using it in this way compresses all those blanks to nothing.

I particularly like to use this with %PUT statements when I'm putting out a message in the log as shown in Program 4.13 and SAS Log 4.12.

Program 4.13: Using the %CMPRES() Function with a %PUT Statement

```
%put ERR%str(OR:) This is a %cmpres(
                  ) message;
```

SAS Log 4.12: Using the %CMPRES() Function with a %PUT Statement

```
ERROR: This is a message
```

Using the IN Operator

Have you ever had to check to see whether a macro variable is equal to more than one value? You probably would use code something like this:

```
%if (&var eq 1 or &var eq 3 or &var eq 5) %then
```

This works, but it can be a lot of typing. In a DATA step, you can use the IN operator:

```
if (var in (1, 3, 5)) then
```

It is a little obscure, but in SAS 9, you can use the IN operator in a %IF macro statement. You must specify a couple of options to turn on this functionality. You can use an OPTIONS statement:

```
options minoperator mindelimiter = ',';
```

You can also use the %MACRO statement:

```
%macro in_test / minoperator mindelimiter = ',';
```

In whichever place you specify them, the options function as follows:

- The MINOPERATOR option turns on the ability to use the IN operator in macro code.
- The MINDELIMITER option defines the delimiter that you want to use between the values.

If you are setting these options on an OPTIONS statement, they must be set before the macro is defined.

The example in Program 4.14 and SAS Log 4.13 uses * as the delimiter between the values, and it also uses the alias, #, for the IN operator.

Program 4.14: Using the IN Operator with a * Delimiter

```
%macro in_test / minoperator mindelimiter = '*';

   %let var = 3;
   %if (&var in 1*3*5) %then
      %put ODD &=var;
   %if (&var # 2*4*6) %then
      %put EVEN &=var;

%mend;
%in_test;
```

SAS Log 4.13: Using the IN operator with a * Delimiter

```
ODD VAR=3
```

I find that version difficult to read, so I prefer to simulate the syntax of the DATA step IN operator: a comma delimiter, the IN operator, and parentheses around the list of values. Program 4.15 and SAS Log 4.14 show how I code the IN operator in a macro.

Program 4.15: Using the IN operator with DATA Step-Style Syntax

```
%macro in_test / minoperator mindelimiter = ',';

    %let var = 3;
    %if (&var in (1, 3, 5)) %then
        %put ODD &=var;

%mend;
%in_test;
```

SAS Log 4.14: Using the IN operator with DATA Step-Style Syntax

```
ODD VAR=3
```

If you need to check whether the value is not in the list, use this syntax:

```
%if (not(&var in (1, 3, 5))) %then
```

It is invalid to put the NOT immediately before the IN.

Testing for Blanks

Checking to see whether a macro variable is blank is a pretty common thing to do in a macro. I've seen all sorts of syntax for doing this, such as the following:

```
%if &var eq %then
%if &var eq %str( ) %then
%if "&var" eq "" %then
```

All of these work. You just need to choose a syntax that you find readable. My preference is to put parentheses around the condition and have no value on the right side of the operator. I think doing this makes the code easy to read:

```
%if (&var eq ) %then
```

I don't like using %STR( ) or quoting everything. It's not necessary, and I don't like extra typing unless it makes the code more legible.

Validating Parameters

When I write a macro that might be used in different places and by different people, I try to be very diligent about checking the parameters for valid values and returning appropriate messages and return codes rather than just allowing the program to fail with syntax errors.

Program 4.16 and SAS Log 4.15 show an example of validating the parameters in a macro. The value for the param parameter is supposed to be Y or N, but I make it easy for the calling program by upcasing the value and keeping only the first character before I validate. This allows the caller to send "yes" and not get an error because of it.

Program 4.16: Validating Parameters

```
%macro parameter_test(param=) / minoperator mindelimiter=',';

   /* check if value is blank */
   %if (%bquote(&param) eq ) %then
   %do;
      %put ERR%str(OR:) The param parameter is required.;
      %return;
   %end;

   /* upcase and take off just the first letter */
   %let param = %upcase(%substr(%bquote(&param), 1, 1));

   /* make sure that it is Y or N */
   %if (not (%bquote(&param) in (N, Y))) %then
   %do;
      %put ERR%str(OR:) The param parameter must be N or Y.;
      %return;
   %end;

   %put &=param;

%mend parameter_test;

%parameter_test(param = Y);
%parameter_test(param = yes);
%parameter_test;
%parameter_test(param = X-Y);
```

SAS Log 4.15: Validating Parameters

```
24  %parameter_test(param = Y);
PARAM=Y
25  %parameter_test(param = yes);
PARAM=Y
26  %parameter_test;
ERROR: The param parameter is required.
27  %parameter_test(param = X-Y);
ERROR: The param parameter must be N or Y.
```

If I have any parameters that have default values, I specify the default value on the %MACRO statement, but I also test for a blank and reset the macro variable to the default in case the calling program sends a blank that overwrites the default value. Program 4.17 and SAS log 4.16 show the test for default values.

Program 4.17: Setting Default Values

```
%macro parameter_test(param = ABC);

   %if (%bquote(&param) eq ) %then
      %let param = ABC;

   %put &=param;

%mend parameter_test;

%parameter_test(param = Y);
%parameter_test;
%parameter_test(param = );
```

SAS Log 4.16: Setting Default Values

```
10   %parameter_test(param = Y);
PARAM=Y
11   %parameter_test;
PARAM=ABC
12   %parameter_test(param = );
PARAM=ABC
```

I generally use the %BQUOTE() function when testing the values in case anyone sends in a strange value that would cause the macro compiler to be confused. For example, if the macro is called with a parameter value of "X-Y", and you don't use %BQUOTE() around the value, the macro compiler attempts to subtract Y from X. This is not possible, so you get an error. I've found the %BQUOTE() function handles quoting most of values that might be sent as parameters.

Creating Macro Functions

SAS macros can be written as traditional macros that do some processing. They can also be written as functions that return a value to the calling program.

To return a value, you just specify the value in the open code of the macro. For example, Program 4.18 and SAS Log 4.17 show a macro that calculates the square of the parameter. The value is returned to the calling program just by putting the value (in the macro variable &sqr) in the open code of the macro without a semicolon.

Program 4.18: Creating and Using a Macro Function

```
%macro square(num);

   %let sqr = %eval(&num ** 2);

   /* return value - no semi-colon */
   &sqr

%mend square;

%let a = %square(2);
%put &=a;
```

```
data _null_;
   a = %square(3);
   putlog a=;
   a = %square(25);
   putlog a=;
run;
```

SAS Log 4.17: Creating and Using a Macro Function

```
11   %put &=a;
A=4
12
13   data _null_;
14      a = %square(3);
15      putlog a=;
16      a = %square(25);
17      putlog a=;
18 run;

a=9
a=625
```

Macro functions are very useful, but have several limitations. They can contain only macro statements. They can't contain any DATA steps, procedures, options, or other SAS statements such as LIBNAME or OPTIONS statements. If you need to run a DATA step to do some processing, you cannot create a macro function; you must create a standard macro. If you must use a standard macro, you can return a value to the calling program by creating a global macro variable, as follows:

```
%global outvar;
%let outvar = &var;
```

The calling program can then use the global variable to retrieve the value.

Chapter 5: SAS Programming

Introduction

This chapter contains a somewhat random collection of SAS programming tips and techniques to help you make your SAS programs as efficient as possible.

Using the ABORT Statement

The ABORT statement can be used from inside a DATA step (ABORT), from within a macro (%ABORT) or from open code (%ABORT). It does the same thing wherever you use it: The program stops executing.

The ABORT statement and %ABORT have several arguments that have slightly different outcomes: ABEND, RETURN, and CANCEL. They can also be used with no arguments. I have found that ABORT CANCEL works best in every situation.

When run from a DATA step, the ABORT statement with any of the arguments has the same basic result: The DATA step processing stops right away and no data is saved to the output data set. The %ABORT and ABORT statements also have the following results, depending on which argument you choose:

ABORT

> Batch and non-interactive mode: The program goes into syntax check mode and might do some macro processing.

> Interactive mode: An error message is written to the log, but the rest of the program continues to run.

ABORT ABEND

> Batch and non-interactive mode: The statement stops all processing of the program.

> Interactive mode: The statement stops processing of the program and closes the interactive SAS session.

ABORT RETURN

> Batch and non-interactive mode: The statement stops all processing of the program.

> Interactive mode: The statement stops processing of the program and closes the interactive SAS session.

ABORT CANCEL

> Batch and non-interactive mode: The statement stops all processing of the program.

> Interactive mode: The statement stops processing of the currently submitted code, but the session does not close and subsequent submissions are not affected.

> Workspace server or stored process: The statement stops processing of the currently submitted code, but subsequent submissions are not affected.

The ABORT CANCEL statement is my favorite version because I can test my programs in interactive SAS and then run them in batch, and I get the same results. The ABORT ABEND and ABORT RETURN statements have often caused me problems by closing my interactive SAS session when I wasn't expecting it, so I never use them. I think that ABORT without arguments isn't severe enough, because if I'm going to abort, I don't want any subsequent code to run.

Updating Option Values

SAS system options are easy to set. You just specify the values in an OPTIONS statement like this:

```
options mprint bufno = max;
```

But what about the options that can contain a list of values, such as FMTSEARCH, SASAUTOS, AUTOEXEC, CMPLIB, HELPLOC, MAPS, MSG, SASHELP, SASSCRIPT, and SET? If you want to maintain the existing list and just add an additional value to the beginning or end of the list, you can use the INSERT or APPEND option.

The INSERT option adds the new value to the beginning of the list, as shown in Program 5.1 and SAS Log 5.1.

Program 5.1: Inserting Values into an Option String

```
%let fmtsearch = %sysfunc(getoption(fmtsearch));
%put &=fmtsearch;
options insert = (fmtsearch = formats);
%let fmtsearch = %sysfunc(getoption(fmtsearch));
%put &=fmtsearch;
```

SAS Log 5.1: Inserting Values into an Option String

```
1    %let fmtsearch = %sysfunc(getoption(fmtsearch));
2    %put &=fmtsearch;
FMTSEARCH=(WORK LIBRARY)
3    options insert = (fmtsearch = formats);
4    %let fmtsearch = %sysfunc(getoption(fmtsearch));
5    %put &=fmtsearch;
FMTSEARCH=(FORMATS WORK LIBRARY)
```

The APPEND option adds the new value to the end of the list, as shown in Program 5.2 and SAS Log 5.2.

Program 5.2: Appending Values to an Option String

```
%let fmtsearch = %sysfunc(getoption(fmtsearch));
%put &=fmtsearch;
options append = (fmtsearch = formats);
%let fmtsearch = %sysfunc(getoption(fmtsearch));
%put &=fmtsearch;
```

SAS Log 5.2: Appending Values to an Option String

```
1    %let fmtsearch = %sysfunc(getoption(fmtsearch));
2    %put &=fmtsearch;
FMTSEARCH=(WORK LIBRARY)
3    options append = (fmtsearch = formats);
4    %let fmtsearch = %sysfunc(getoption(fmtsearch));
5    %put &=fmtsearch;
FMTSEARCH=(WORK LIBRARY FORMATS)
```

Getting Information from SASHELP Views

The SASHELP library is full of data, views, and catalogs. Some of these are used to keep SAS working, some are samples, and some—in particular, the views—are available for you to use in your SAS.

The views contain information about your SAS session, including what tables are available, what macros and macro variables have been defined, and what options are set. The view names all begin with "v", so they are easy to locate in the SASHELP library.

One thing to keep in mind with these views is that the larger your SAS session is, the slower the views are. So if you have many data tables with many variables, and you have many macros and macro variables defined, performance might not be acceptable. In that case, you might need to use another method to get this information.

Here is a small subset of views that I find most useful:

sashelp.vtable

> The vtable view gives header information about each table that is available in the system. For every libref that you have, the view contains one record for each table.

sashelp.vcolumn

> The vcolumn view contains a row for every variable (column) in every table, including the name, type, size, format, and label, as well as plenty of other information.

sashelp.vindex

> The vindex view has all the information about any indexes that have been created on your tables, with one record for each index.

sashelp.vformat

> The vformat view has one record for each format and informat available to the SAS session, including the formats supplied by SAS as well as the user-defined formats.

sashelp.vcatlg

> The vcatlg view contains a record for every catalog member in all the catalogs available to the SAS session. These records include all the catalogs provided in the SASHELP library as well as any user-created catalogs.

sashelp.vextfl

> The vextfl view contains all the external files that have been defined to the system. If the fileref begins with #LN, then this is a temporary filename that is created by SAS, or it is syntax to point to a temporary file when you use the filename x temp.

sashelp.vmacro

> The vmacro view contains all the macro variables defined to the system, including their scope (AUTOMATIC, GLOBAL, LOCAL) and their value. This view is useful when looking for macro variables with a certain naming convention.

Creating a Unique Key with PROC SQL

It is simple to use a DATA step to add a variable to a data table that contains a unique key for each row. Just create a variable and add one to it each time you write the record. Achieving the same result with PROC SQL is just as easy, but not obvious: Use the MONOTONIC function. This function enables you to create a unique numeric value by keeping a counter and putting the number on each row that is processed, so the result is a unique key in your data table. See the example in Program 5.3 to see how to create a simple unique key. Figure 5.1 shows the output table created by PROC SQL.

Program 5.3: Adding a Unique Key

```
proc sql;
   create table class as
      select monotonic() as key,
            *
         from sashelp.class
            where age gt 13;
quit;
```

Figure 5.1: Adding a Unique Key

Obs	key	Name	Sex	Age	Height	Weight
1	1	Alfred	M	14	69.0	112.5
2	2	Carol	F	14	62.8	102.5
3	3	Henry	M	14	63.5	102.5
4	4	Janet	F	15	62.5	112.5
5	5	Judy	F	14	64.3	90.0
6	6	Mary	F	15	66.5	112.0
7	7	Philip	M	16	72.0	150.0
8	8	Ronald	M	15	67.0	133.0
9	9	William	M	15	66.5	112.0

However, because PROC SQL doesn't necessarily process the rows in the order you might expect, especially when joining tables, the key might not correspond to the row number. If you need the key and the row number to match, then you need to use a DATA step instead of PROC SQL. The example in Program 5.4 shows a join, and you can see in the output table in Figure 5.2 that the KEY variable created by the MONOTONIC function does not contain consecutive numbers.

Program 5.4: Adding a Unique Key in a Join

```
proc sql;
   create table country as
      select monotonic() as key,
             d.name, d.id, d.region, f.file
         from sashelp.flags as f
               inner join
               sashelp.demographics as d
                  on f.title eq d.name or
                     f.title eq d.isoname
            where substr(f.title, 1, 1) eq "A";
quit;
```

Figure 5.2: Adding a Unique Key in a Join

Obs	key	NAME	ID	region	FILE
1	332	AFGHANISTAN	110	EMR	AFGHANIS
2	431	ALBANIA	120	EUR	ALBANIA
3	673	ALGERIA	125	AFR	ALGERIA
4	1024	ANDORRA	140	EUR	ANDORRA
5	1265	ANGOLA	141	AFR	ANGOLA
6	1790	ANTIGUA/BARBUDA	149	AMR	ANTIGUAB
7	1988	ARGENTINA	150	AMR	ARGENTIN
8	2205	ARMENIA	135	EUR	ARMENIA
9	2744	ASHMORE/CARTIER	155	WPR	AUSTRALI
10	2745	AUSTRALIA	160	WPR	AUSTRALI
11	2748	CORAL SEA ISLANDS	294	WPR	AUSTRALI
12	2798	AUSTRIA	165	EUR	AUSTRIA
13	2991	AZERBAIJAN	115	EUR	AZERBAIJ

Setting a Boolean

SAS doesn't have a Boolean type of variable, so we rely on a numeric value that is 0 for false or any other value for true. So if you want to use a variable as a Boolean, you need to set it to a 0 or another number. You can set it like this:

```
if (sex eq "M") then
    male = 1;
else
    male = 0;
```

As an alternative, the example in Program 5.5 and SAS Log 5.3 shows that you can put a condition on the right side of the equal sign, and the condition is evaluated to either a 0 or 1, which is then assigned to the variable. The parentheses around the expression are not needed, but I think they make it easier to read.

Program 5.5: Setting a Boolean

```
data _null_;
    set sashelp.class;
    male = (sex eq "M");
    putlog sex= male=;
run;
```

SAS Log 5.3: Setting a Boolean

```
Sex=M male=1
Sex=F male=0
Sex=F male=0
Sex=F male=0
Sex=M male=1
. . .
```

The results from the INDEX() function can be used as a Boolean because the function returns a 0 if it doesn't find the substring (0, so false), and the position of the substring if it does find it (greater than 0, so true). Program 5.6 and SAS Log 5.4 show how to use the INDEX() function as a Boolean, rather than using a comparison to create the Boolean value.

Program 5.6: Using the INDEX() Function as a Boolean

```
data _null_;
   set sashelp.class;
   if (index(upcase(name), "H")) then
      putlog name=;
run;
```

SAS Log 5.4: Using the INDEX() Function as a Boolean

```
Name=Henry
Name=John
Name=Philip
Name=Thomas
```

Accumulating Values

Often you need to create a counter or to accumulate a value. For example, you can calculate an average by keeping a count of the number of values and running sum of all the values, and then dividing the sum by the number to get the average. You could use code in a DATA step like this:

```
retain totalAge 0;
retain numStudents 0;
totalAge = totalAge + age;
numStudents = numStudents + 1;
if (eof) then
   averageAge = totalAge / numStudents;
```

Or you can use the SAS sum statement:

```
var + value;
```

This syntax does the accumulation and also implicitly initializes the accumulator variable to 0 and retains its value over iterations of the DATA step. This means that you can simplify the code to look like the example in Program 5.7 and SAS Log 5.5.

Program 5.7: Accumulating Values

```
data _null_;
   set sashelp.class end = eof;
   totalAge + age;
   numStudents + 1;
   if (eof) then
   do;
      averageAge = totalAge / numStudents;
      putlog numStudents= totalAge= averageAge=;
   end;
run;
```

SAS Log 5.5: Accumulating Values

```
numStudents=19 totalAge=253 averageAge=13.315789474
```

It is important that the accumulator variable is a new variable created by the DATA step, and not a variable that is in the input table. If the accumulator is in the incoming data, then the value from the input table or file replaces the accumulated value each time the data is read.

Replacing a Substring

The SUBSTR() function is pretty basic. It enables you to get a substring from a character string using a start point and the length of the substring. Did you know that you can also use the SUBSTR() function on the left side of the equal sign in an assignment statement? When you do this, you can assign the section of the string specified by the start point and length to another string. Program 5.8 ad SAS Log 5.6 show an example of replacing a substring.

Program 5.8: Replacing a Substring

```
data _null_;
   set sashelp.class;
   string = "/*           */";
   substr(string, 4, length(name)) = name;
   putlog string=;
run;
```

SAS Log 5.6: Replacing a Substring

```
string=/* Alfred    */
string=/* Alice     */
string=/* Barbara   */
string=/* Carol     */
string=/* Henry     */
string=/* James     */
string=/* Jane      */
...
```

This is not a syntax that you will need too often. I have used it to create output that lines up nicely, like the preceding example where I wanted the */ at the end of each line to be in the same position in each string. To do this without the SUBSTR() function on the left of the equal sign, I would have to check lengths of the variables and add the right number of spaces. The SUBSTR() on the left makes this much simpler.

Using Data Values Tables to Create and Run Code

What do you do when you want to run some code that uses values that are in a data table? There are plenty of ways to do this, including creating macro variable arrays from the data, creating SAS code in a temporary file and then using %INCLUDE to run it, or using CALL EXECUTE() to generate and run SAS code.

The following example shows the difference between the three methods:

I want to copy sashelp.class to the WORK library, rename all the variables so that they begin with X_, and add a new load_date variable that contains today's date.

The first step, in Program 5.9, is to use PROC CONTENTS to get a list of the variables and their attributes from the original table.

Program 5.9: Getting Variables and Attributes from the Original Table

```
proc contents data = sashelp.class
               out = contents
               noprint;
run;
```

Using Macro Variable Arrays

Simulating a macro variable array (&var1, &var2, and so on) is very easy with either PROC SQL or the DATA step.

The PROC SQL code in Program 5.10 creates a macro variable for each row of the contents table. The sashelp.class table contains 5 variables, so this code creates five macro variables: &varName1, &varName2, … &varName5. The automatic macro variable &SQLOBS contains the number of records that are read, so in this case it coincides with the number of macro variables that are created. Note that even though the code has 9999 macro variables listed in the INTO clause, PROC SQL is smart and only creates the number of macro variables that are needed. So in this case, the macro variables &varName6 through &varName9999 will not exist after running this code. SAS Log 5.7 shows the macro variables that are created by PROC SQL.

Program 5.10: Creating a Macro Array with PROC SQL

```
proc sql noprint;
    select name into :varName1-:varName9999
        from contents;
quit;
%let numVars = &sqlobs;

%put &=numVars;
%put &=varName1;
%put &=varName2;
%put &=varName3;
%put &=varName4;
%put &=varName5;
```

SAS Log 5.7: Creating a Macro Array with PROC SQL

```
7    %put &=numVars;
NUMVARS=5
8    %put &=varName1;
VARNAME1=Age
9    %put &=varName2;
VARNAME2=Height
10   %put &=varName3;
VARNAME3=Name
11   %put &=varName4;
```

```
VARNAME4=Sex
12   %put &=varName5;
VARNAME5=Weight
```

Tip: If you want to make more than one macro array with PROC SQL, this is the syntax:

```
proc sql noprint;
   select name, label
           into :varName1-:varName9999,
                :varLabel1-:vaLabel9999
      from contents;
quit;
```

Tip: Be sure to include the noprint option on PROC SQL when you are not creating an output table. If you leave this off, then PROC SQL will create a report of the data.

The DATA step code in Program 5.11 produces the same results as PROC SQL – it creates the five macro variables, and it puts the number of macro variables in &numVars. Be sure to initialize &numVars to 0 before the DATA step; if there is no data in the contents table, then &numVars will not exist and this might cause issues in the rest of your code. SAS Log 5.8 shows the macro variables that were created by the DATA step.

Program 5.11: Creating a Macro Array with a DATA Step

```
%let numVars = 0;
data _null_;
   set contents end = eof;
   call symputx(cats("varName", _n_), name);
   if (eof) then
      call symputx("numVars", _n_);
run;

%put &=numVars;
%put &=varName1;
%put &=varName2;
%put &=varName3;
%put &=varName4;
%put &=varName5;
```

SAS Log 5.8: Creating a Macro Array with a DATA Step

```
9    %put &=numVars;
NUMVARS=5
10   %put &=varName1;
VARNAME1=Age
11   %put &=varName2;
VARNAME2=Height
12   %put &=varName3;
VARNAME3=Name
13   %put &=varName4;
VARNAME4=Sex
```

```
14   %put &=varName5;
VARNAME5=Weight
```

After the macro variable array has been created, it can be used to generate code. The macro in Program 5.12 uses a %DO loop to generate the RENAME statements to rename all the variables in the new work table. SAS Log 5.9 shows the SAS code that is generated by the macro.

Program 5.12: Macro to Create a New Table

```
options mprint;
%macro makeTable;
   data work.class;
      set sashelp.class
          (rename = (
              %do i = 1 %to &numVars;
                  &&varName&i = X_&&varName&i
              %end;
          ));
      load_date = today();
      format load_date date9.;
   run;
%mend makeTable;
%makeTable;
```

SAS Log 5.9: Macro to Create a New Table

```
MPRINT(MAKETABLE):   data work.class;
MPRINT(MAKETABLE):   set sashelp.class (rename = ( Age = X_Age
Height = X_Height Name = X_Name Sex = X_Sex Weight = X_Weight ));
MPRINT(MAKETABLE):   load_date = today();
MPRINT(MAKETABLE):   format load_date date9.;
MPRINT(MAKETABLE):   run;
```

Creating and Including Code

Another method of running SAS code on the basis of data values is to create the code, store it in a temporary file, and use the %INCLUDE statement to run it. You can use a SAS DATA step with a FILE statement and PUT statements to create the code, which means that you have all the power of the DATA step to manipulate the data as much as you want. Use a FILENAME statement with the temp device type to create a temporary file. You don't have to specify a name, and the file is deleted when the filename is cleared. Program 5.13 and SAS Log 5.10 show how to create the SAS code and then run it.

Program 5.13: Using a Temporary File to Create and Run a DATA Step

```
filename code temp;

data _null_;
   set contents;
   by memname;
   file code;
   if (first.memname) then
   do;
      put "data work." memname +(-1) ";";
      put "   set sashelp." memname;
      put "      (rename = (";
   end;
   put "             " name "= X_" name;
   if (last.memname) then
   do;
      put "      ));";
      put "   load_date = today();";
      put "   format load_date date9.;";
      put "run;";
   end;
run;

options source2;
%include code;

filename code;
```

SAS Log 5.10: Using a Temporary File to Create and Run a DATA Step

```
24  %include code;
NOTE: %INCLUDE (level 1) file CODE is file C:\Temp\SAS Temporary
Files\_TD12196_L7A678_\#LN00117.
25 +data work.CLASS;
26 +   set sashelp.CLASS
27 +         (rename = (
28 +                   Age = Y_Age
29 +                   Height = Y_Height
30 +                   Name = Y_Name
31 +                   Sex = Y_Sex
32 +                   Weight = Y_Weight
33 +         ));
34 +   load_date = today();
35 +   format load_date date9.;
36 +run;
```

Using CALL EXECUTE

Very similar to the %INCLUDE method is the CALL EXECUTE() routine. I tend to use this method more often than %INCLUDE because it is a little more flexible and enables me to use functions when creating lines of SAS code.

The CALL EXECUTE() routine accumulates the SAS statements in a buffer. As soon as the DATA step is finished, the SAS software runs the statements for you. Be careful with macros in CALL EXECUTE(), because they might execute or resolve before you expect them to. Use single quotation marks around your macro code if you want the macro to execute when the CALL EXECUTE() code is running. Program 5.14 and SAS Log 5.11 show an example of using CALL EXECUTE() to generate and run SAS code.

Program 5.14: Using CALL EXECUTE() to Generate and Run SAS Code

```
data _null_;
   set contents;
   by memname;
   if (first.memname) then
   do;
      call execute("data work.Y_" !! strip(memname) !! ";");
      call execute("   set sashelp." !! strip(memname));
      call execute("      (rename = (");
   end;
   call execute("             " !!
             strip(name) !! " = Y_" !! strip(name));
   if (last.memname) then
   do;
      call execute("      ));");
      call execute("  load_date = today();");
      call execute("  format load_date date9.;");
      call execute("run;");
   end;
run;
```

SAS Log 5.11: Using CALL EXECUTE() to Generate and Run SAS Code

```
NOTE: CALL EXECUTE generated line.
1    + data work.Y_CLASS;
2    +    set sashelp.CLASS
3    +          (rename = (
4    +                      Age = Y_Age
5    +                      Height = Y_Height
6    +                      Name = Y_Name
7    +                      Sex = Y_Sex
8    +                      Weight = Y_Weight
9    +          ));
10   +    load_date = today();
11   +    format load_date date9.;
12   + run;
```

Taking Control of DATA Step Processing

The DATA step is an incredibly powerful tool that has many built-in features, including the implicit loop that runs the code for each record read from an incoming data table or file. Much of the time it is sufficient to let this implicit loop do its job, but there are times that you want to move through the code differently.A number of statements and features enable you to do this. You might be aware of some or all of these statements, but it might help to see them all defined together so that you can choose the appropriate method for your situation.

Branching

Branching to another section of code using the GOTO or LINK statements is an old-fashioned way of programming, and most programmers will tell you not to do it. I agree; if you can find another way to branch, please do. Both of these statements are confusing to read and follow for anyone maintaining your program, including you. That being said, there are rare cases when you might want to use these statements, such as branching to a section of code that handles error conditions.

Using GOTO

When you use a GOTO statement, the program immediately branches to the specified label and continues from there until it reaches a RETURN statement or the end of the DATA step. If you want to use the GOTO statement to branch to an error condition, you could do something like the example in Program 5.15 and SAS Log 5.12.

Program 5.15: Using GOTO to Branch to an Error Condition

```
data _null_;
   set sashelp.class;
   if (age gt 15) then
   do;
      msg = "greater than 15";
      goto err;
   end;
   else if (age lt 12) then
   do;
      msg = "less than 12";
      goto err;
   end;
   put "Good record: " age= name=;
   return;
   err:
      put "ERR" "OR: Age is " msg +(-1) ": "
         age= name=;
run;
```

SAS log 5.12: Using GOTO to Branch to an Error Condition

```
Good record: Age=14 Name=Alfred
Good record: Age=13 Name=Alice
Good record: Age=13 Name=Barbara
Good record: Age=14 Name=Carol
Good record: Age=14 Name=Henry
Good record: Age=12 Name=James
Good record: Age=12 Name=Jane
Good record: Age=15 Name=Janet
Good record: Age=13 Name=Jeffrey
Good record: Age=12 Name=John
ERROR: Age is less than 12: Age=11 Name=Joyce
Good record: Age=14 Name=Judy
Good record: Age=12 Name=Louise
Good record: Age=15 Name=Mary
ERROR: Age is greater than 15: Age=16 Name=Philip
Good record: Age=12 Name=Robert
Good record: Age=15 Name=Ronald
```

```
ERROR: Age is less than 12: Age=11 Name=Thomas
Good record: Age=15 Name=William
```

Note that it is necessary to include the RETURN statement before the err: label. This is because the processing keeps going even when it reaches a label, so the RETURN statement is necessary to send the processing back to the beginning of the DATA step.

I still prefer other methods than a GOTO statement in this case. I think a macro or an IF statement with a RETURN statement is easier to read, debug, and maintain.

Tip: You can also use this style of error processing in macros as well as the DATA step. In a macro there is a %GOTO statement and a %RETURN statement that can be used in the same manner. Other styles are still easier to read and maintain.

Using LINK

The LINK statement is similar to the GOTO statement, and I feel even more strongly that a different style of programming is preferable to using a LINK statement. The difference between LINK and GOTO is that if the labeled section that you link to has a RETURN statement at the end, you are returned to the next statement after the LINK statement instead of the top of the DATA step. It makes the code very difficult to follow, especially if you have GOTO statements and LINK statements in the same DATA step branching to the same labeled section. The maintenance is incredibly difficult.

Program 5.16 and SAS Log 5.13 are an example of using the LINK statement.

Program 5.16: Using a LINK Statement

```
data _null_;
   set sashelp.class;
   errFlag = 0;
   length msg $25;
   if (age gt 15) then
   do;
      msg = "Age greater than 15";
      link err;
   end;
   if (height ge 66) then
   do;
      msg = "Height greater than 66";
      link err;
   end;
   putlog "Record: " age= name= height= errflag=;
   return;
 err:
   putlog "ERR" "OR: " msg;
   errFlag = 1;
   return;
run;
```

SAS Log 5.13: Using a LINK Statement

```
ERROR: Height greater than 66
Record: Age=14 Name=Alfred Height=69 errFlag=1
Record: Age=13 Name=Alice Height=56.5 errFlag=0
Record: Age=13 Name=Barbara Height=65.3 errFlag=0
Record: Age=14 Name=Carol Height=62.8 errFlag=0
Record: Age=14 Name=Henry Height=63.5 errFlag=0
Record: Age=12 Name=James Height=57.3 errFlag=0
Record: Age=12 Name=Jane Height=59.8 errFlag=0
Record: Age=15 Name=Janet Height=62.5 errFlag=0
Record: Age=13 Name=Jeffrey Height=62.5 errFlag=0
Record: Age=12 Name=John Height=59 errFlag=0
Record: Age=11 Name=Joyce Height=51.3 errFlag=0
Record: Age=14 Name=Judy Height=64.3 errFlag=0
Record: Age=12 Name=Louise Height=56.3 errFlag=0
ERROR: Height greater than 66
Record: Age=15 Name=Mary Height=66.5 errFlag=1
ERROR: Age greater than 15
ERROR: Height greater than 66
Record: Age=16 Name=Philip Height=72 errFlag=1
Record: Age=12 Name=Robert Height=64.8 errFlag=0
ERROR: Height greater than 66
Record: Age=15 Name=Ronald Height=67 errFlag=1
Record: Age=11 Name=Thomas Height=57.5 errFlag=0
ERROR: Height greater than 66
Record: Age=15 Name=William Height=66.5 errFlag=1
```

Just because you can do it doesn't mean you should do it!

Returning

The RETURN statement is very useful in contexts other than the GOTO and LINK statements, particularly in an IF statement. It stops processing of the current iteration of the DATA step and returns to the top of the DATA step. If there is an implicit output (that is, there is no output statement in the DATA step), the current record is saved to the output data table.

> **Tip:** Every DATA step has an implicit OUTPUT and an implicit RETURN statement at the end. This means that at the end of the DATA step, SAS saves the current data to the output table and then returns to the beginning of the DATA step to process the next record. If there is an actual output statement in the DATA step, then the implicit output is not used.

In the example in Program 5.17, the DATA step reads the sashelp.class table. When the value of age is less than 13, the type variable is not set to anything, because the RETURN statement sends the processing back to the top of the DATA step after writing the record. Figure 5.3 shows the output created by the DATA step.

Program 5.17: Using a RETURN Statement

```
data table (keep = name age type);
   set sashelp.class;
   length type $15;
   if (age lt 13) then
       return;
   type = "Teenager";
run;
```

Figure 5.3: Using a RETURN Statement

Obs	Name	Age	type
1	Alfred	14	Teenager
2	Alice	13	Teenager
3	Barbara	13	Teenager
4	Carol	14	Teenager
5	Henry	14	Teenager
6	James	12	
7	Jane	12	
8	Janet	15	Teenager
9	Jeffrey	13	Teenager
10	John	12	
11	Joyce	11	
12	Judy	14	Teenager
13	Louise	12	
14	Mary	15	Teenager
15	Philip	16	Teenager
16	Robert	12	
17	Ronald	15	Teenager
18	Thomas	11	
19	William	15	Teenager

Deleting

The DELETE statement is like the RETURN statement in that it returns processing to the top of the DATA step. The difference is that it does not save the current record to the output data table; the implicit output is not executed.

Program 5.18 uses the same example as Progam 5.19. Anyone whose age is less than 13 is not saved to the output data table. Figure 5.4 shows the output from the DATA step.

Program 5.18: Using a DELETE Statement

```
data table (keep = name age type);
   set sashelp.class;
   length type $15;
   if (age lt 13) then
      delete;
   type = "Teenager";
run;
```

Figure 5.4: Using a DELETE Statement

Obs	Name	Age	type
1	Alfred	14	Teenager
2	Alice	13	Teenager
3	Barbara	13	Teenager
4	Carol	14	Teenager
5	Henry	14	Teenager
6	Janet	15	Teenager
7	Jeffrey	13	Teenager
8	Judy	14	Teenager
9	Mary	15	Teenager
10	Philip	16	Teenager
11	Ronald	15	Teenager
12	William	15	Teenager

Subsetting

The subsetting IF statement is similar to an IF statement with a DELETE statement, except that it stops processing of the current iteration of the DATA step, does not save the record, and returns to the beginning of the DATA step if the condition is false rather than deleting the record when the condition is true. A subsetting IF statement does not have a THEN part of the statement. It just consists of IF and a condition. Recycling the examples in Program 5.17 and Program 5.18, Program 5.19 shows how the subsetting IF statement works. The output from the DATA step is in Figure 5.5.

Program 5.19: Using a Subsetting IF Statement

```
data table (keep = name age type);
   set sashelp.class;
   length type $15;
   if (age ge 13);
   type = "Teenager";
run;
```

Figure 5.5: Using a Subsetting IF Statement

Obs	Name	Age	type
1	Alfred	14	Teenager
2	Alice	13	Teenager
3	Barbara	13	Teenager
4	Carol	14	Teenager
5	Henry	14	Teenager
6	Janet	15	Teenager
7	Jeffrey	13	Teenager
8	Judy	14	Teenager
9	Mary	15	Teenager
10	Philip	16	Teenager
11	Ronald	15	Teenager
12	William	15	Teenager

Note that the condition of the IF statement was changed in order to keep all the records that have an age greater than or equal to 13.

Stopping

The STOP statement will let you stop all processing of the DATA step as soon as the STOP statement is executed. The output table will be created with the records that were added until the STOP was executed. The data that is being processed in the current iteration of the DATA step is not saved to the data table unless there is an explicit output statement executed before the STOP statement.

The STOP statement is very important when you take control of the input data using something like the SET statement and the POINT= or the KEY= options. When you do this, the DATA step no longer knows when to stop processing, so you need to use the STOP statement to avoid getting into an infinite loop.

In the example in Program 5.20 and SAS Log 5.14, the processing stops when the name is equal to "Henry". The next record after "Carol" is "Henry", but the STOP statement kept the "Henry" record from being written to the table and stopped the DATA step processing. The output from the DATA step is in Figure 5.6.

Program 5.20: Using the STOP Statement

```
data table (keep = name age);
   set sashelp.class;
   putlog "Before Stop: " name=;
   if (name eq "Henry") then
      stop;
   putlog "After Stop: " name=;
run;
```

SAS Log 5.14: Using the STOP Statement

```
Before Stop: Name=Alfred
After Stop: Name=Alfred
Before Stop: Name=Alice
After Stop: Name=Alice
Before Stop: Name=Barbara
After Stop: Name=Barbara
Before Stop: Name=Carol
After Stop: Name=Carol
Before Stop: Name=Henry
```

Figure 5.6 Using the STOP Statement

Obs	Name	Age
1	Alfred	14
2	Alice	13
3	Barbara	13
4	Carol	14

Aborting

As the name implies, the ABORT statement is very final. It stops processing of the DATA step as soon as it is executed, and the output table that was being created is not saved. See the "Using the ABORT Statement" section for more details.

Continuing

The CONTINUE statement is used in a DO loop to stop processing of the current iteration of the loop, return to the beginning of the loop, and start processing of the next iteration.

In the example in Program 5.21 and SAS Log 5.15, when i equals 3, the CONTINUE statement sends the processing back to the beginning of the loop, so that the second PUTLOG statement is not executed.

Program 5.21: Using the CONTINUE Statement

```
data _null_;
   do i = 1 to 5;
      putlog "Before: " i=;
      if (i eq 3) then
         continue;
      putlog "After: " i=;
   end;
run;
```

SAS Log 5.15: Using the CONTINUE Statement

```
Before: i=1
After: i=1
Before: i=2
After: i=2
Before: i=3
Before: i=4
After: i=4
```

```
Before: i=5
After: i=5
```

Leaving

The LEAVE statement is used in a DO loop. It stops processing of the loop completely and goes to the next statement after the END statement.

In the example in Program 5.22 and SAS Log 5.16, when i equals 3, the LEAVE statement is executed, so the processing of the loop stops and the statement after the END statement is executed.

Program 5.22: Using the LEAVE Statement

```
data _null_;
   do i = 1 to 5;
      putlog "Before: " i=;
      if (i eq 3) then
         leave;
      putlog "After: " i=;
   end;
   putlog "After Do Loop";
run;
```

SAS Log 5.16: Using the LEAVE Statement

```
Before: i=1
After: i=1
Before: i=2
After: i=2
Before: i=3
After Do Loop
```

Warning: Be careful when using any of these statements to change the implicit nature of the DATA step. You might affect other features, including the _N_ variable, the END= variable, the FIRST. and LAST. variables, and the LAG() function. For example, if you use a subsetting IF statement to stop the processing of a record, and that record is the last one in the input table, then any processing that was supposed to be done when you reach the end of the input data table based on the END= variable does not execute.

Figuring Out Where You Are

It is often very important to know where you are when working in the DATA step—what iteration of the DATA step you are on, whether you are at the beginning or ending of the input data, and whether you are at the beginning or ending of a group of values.

Counting Iterations

The automatic variable _N_ is often confused with the number of the record you are reading. It is actually a counter of the number of times the DATA step has iterated. Most of the time, when a data table or raw data file is read, the value of _N_ corresponds to the record number. But it is important to remember that it is really an iteration counter.

Program 5.23 and SAS Log 5.17 show an example of using _N_ to put a title on a simple report during the first iteration and then to put out only the odd-numbered records.

Program 5.23: Using _N_

```
data _null_;
   set sashelp.class;
   if (_n_ eq 1) then
      putlog "This is the class";
   if (mod(_n_, 2) ne 0) then
      putlog @3 _n_ +2 name;
run;
```

SAS Log 5.17: Using _N_

```
This is the class
 1  Alfred
 3  Barbara
 5  Henry
 7  Jane
 9  Jeffrey
11  Joyce
13  Louise
15  Philip
17  Ronald
19  William
```

Tip: If you want the true record number from the table that you are reading, then add the CUROBS= option to the SET statement. You set this option to the name of a variable that is created for you and contains the true record number that was read from the table on the SET statement, even if a WHERE clause has been applied.

Finding the End

To know when you are at the last record of the incoming data, you can use the END= option on the SET, MERGE, UPDATE, MODIFY or INFILE statement. The END= option causes a variable to be created that is set to 0 for every record that is read until the last record, and then it is set to 1.

I usually name my end variable eof (end-of-file). It's easy to remember, rarely conflicts with a variable in my incoming data, is easy to type, and doesn't require me to come up with a creative variable name every time I use END=.

Program 5.24 and SAS Log 5.18 show an example of using the END= option to accumulate a value and then save that value to a macro variable when you are at the end of the DATA step.

Program 5.24: Using the END= Option

```
data _null_;
   set sashelp.class end = eof;
   totHeight + height;
   if (eof) then
   do;
      averageHeight = totHeight / _n_;
      call symputx("averageHeight", averageHeight);
   end;
run;
%put &=averageHeight;
```

SAS Log 5.18: Using the END= Option

```
AVERAGEHEIGHT=62.336842105
```

Finding Group Beginnings and Endings

If you haven't used FIRST.*var* and LAST.*var*, you are missing out on a great tool. The FIRST.*var* automatic variable is created for each variable on the BY statement, and it is set to 1 (true) when the value of *var* is the first record of a BY group. The LAST.*var* automatic variable is also created for all variables on the BY statement, and it is set to 1 (true) when you are at the last record of the current BY group. I think the easiest way to explain FIRST. and LAST. is to look at the example in Program 5.25 and SAS Log 5.19.

Program 5.25: Demonstration of FIRST. and LAST. Processing

```
proc sort data = sashelp.class
          out = class;
   by sex age;
run;

data _null_;
   set class;
   by sex age;
   if (_n_ eq 1) then
      putlog @1 "sex" @5 "age" @9 "first.sex" @19 "last.sex"
             @28 "first.age" @38 "last.age";
   if (first.sex) then
      putlog @1 sex @;
   if (first.age) then
      putlog @5 age @;
   putlog @17 first.sex @25 last.sex
          @35 first.age @44 last.age;
   if (last.sex or last.age) then
      putlog;
run;
```

SAS Log 5.19: Demonstration of FIRST. and LAST. processing

sex	age	first.sex	last.sex	first.age	last.age
F	11	1	0	1	1
	12	0	0	1	0
		0	0	0	1
	13	0	0	1	0
		0	0	0	1
	14	0	0	1	0
		0	0	0	1
	15	0	0	1	0
		0	1	0	1
M	11	1	0	1	1
	12	0	0	1	0
		0	0	0	0
		0	0	0	1
	13	0	0	1	1
	14	0	0	1	0
		0	0	0	1
	15	0	0	1	0
		0	0	0	1
	16	0	1	1	1

In order use FIRST. and LAST., the table must be sorted by one or more variables, and you must include the same BY statement in your DATA step. You can leave off variables from the end of the BY statement if you don't need the FIRST. and LAST. variables to be created.

There are lots of uses for FIRST. and LAST. One example is to help find duplicate records. If the FIRST. and LAST. variables are both 1, then there is only 1 record for this value. Program 5.26 and SAS Log 5.20 show how to search for duplicate records.

Program 5.26: Using FIRST. and LAST. to Find Duplicates

```
proc sort data = sashelp.cars
          out = cars;
   by make type;
run;

data _null_;
   set cars;
   by make type;
   if (first.type eq 0 or last.type eq 0) then
      putlog "Type with duplicates: " type= first.type= last.type=;
   else
      putlog "Type with one value: " type= first.type= last.type=;
run;
```

SAS Log 5.20: Using FIRST. and LAST. to Find Duplicates

```
Type with one value: Make=Acura Type=SUV FIRST.Type=1 LAST.Type=1
Type with duplicates: Make=Acura Type=Sedan FIRST.Type=1 LAST.Type=0
Type with duplicates: Make=Acura Type=Sedan FIRST.Type=0 LAST.Type=0
Type with duplicates: Make=Acura Type=Sedan FIRST.Type=0 LAST.Type=0
Type with duplicates: Make=Acura Type=Sedan FIRST.Type=0 LAST.Type=0
Type with duplicates: Make=Acura Type=Sedan FIRST.Type=0 LAST.Type=1
Type with one value: Make=Acura Type=Sports FIRST.Type=1 LAST.Type=1
Type with duplicates: Make=Audi Type=Sedan FIRST.Type=1 LAST.Type=0
Type with duplicates: Make=Audi Type=Sedan FIRST.Type=0 LAST.Type=0
...
```

Chapter 6: Application Development

Introduction

When you are developing a production application in SAS, there are some additional tasks you should do to ensure that your code is readable and maintainable. Here are some issues to keep in mind that will make you very popular with your co-workers.

Using Comments

Always use comments in your code. You and the person who inherits your code will be thankful for it later.

When I develop code, I add minimal comments. This is because as the code is evolving, comments become obsolete, and obsolete comments are much worse than no comments. As the last step in my development, I go back through the code and add comments. This step forces me to explain each piece of code to myself (think of this as an informal code walk-through). If it is too confusing or I can't explain it, then I rethink that piece of code. Yes, it is an extra step when we are all under deadlines, but it has often enabled me to catch bugs before putting code into production, as well as resulting in fully commented and maintainable code. These are both big time savers in the general life of an application.

Choosing a Style

The following styles of comments are available in SAS:

```
/* */
*  ;
%* ;
```

If you are in a hurry and just want to comment out a line of code, then using the *; style is great, because you just have to type a * at the beginning of the line and you are done. However, there are issues with this style: Macros and macro variables are still evaluated, and you can get mismatched quotation marks. This style can also lead to maintenance problems, such as commenting less than you expected because there are multiple semi-colons on a single line or a semicolon in a string that you didn't notice. You can use the %* ; style to avoid the issue with macros, but you still can have quoting and maintenance issues.

I prefer the /* */ method. This method masks everything between the /* and */, so you don't have to worry about what is in your comment. The /* */ style also enables you to comment out partial statements. For example, if you only want to comment out one variable in a list, you just do this:

```
var A1 A2 /* A3 */ A4;
```

Commenting Out Chunks of Code

How often do you need to comment out a block of code so that you can test something? Putting comment symbols around all the code is a lot of trouble, especially if you already have comments in that section. And you then have to remove the comments when you are finished. So what is the answer? Use a macro. Just add the macro statement %MACRO COMMENT; at the beginning of the block, and add %MEND; at the end of the block. That block won't run because you never call the %COMMENT; macro. Note that you can use any name for the macro. I like to use COMMENT() so that I can easily tell that it is a comment block.

In the example in Program 6.1, the DATA step in the middle of the program is commented out by a macro.

Program 6.1: Commenting Out a Chunk of Code with a Macro

```
proc sort data = sashelp.class
          out = class;
   by name;
run;

%macro comment;
%let x = Jane;
data class;
   set class;
   if (name ne "&x") then
      name = upcase(name);
run;
%mend;
```

```
proc print data = class;
   var name;
run;
```

Dealing with Notes, Warnings, and Errors

I feel strongly that you should get rid of all warning and error messages in production code. Errors are obvious; your program won't run with them. But if you let warnings remain in your code, whoever is maintaining the code will need to waste time checking the messages to make sure that they don't matter. For batch processes that run regularly, I generally run a second program that reads the log file and reports any errors or warnings to the administrator. If there are warnings that don't matter, my parser reports them.

Getting Rid of Common Notes, Warnings, and Errors

Different SAS procedures and the DATA step can produce warning, error, and note messages that indicate when there is a potential issue in processing. You should get rid of these messages from the SAS log so that the program is easier to maintain.

PROC DATASETS Warnings

If you ask PROC DATASETS to do something that it can't do (for example, delete a non-existent data table), it produces a warning. To get rid of the warning, add the NOWARN option to the PROC DATASETS statement:

```
proc datasets lib = work nowarn;
```

> **Tip**: Another nice option for PROC DATASETS is NOLIST. This option stops PROC DATASETS from creating a list in the SAS log of all the tables in the library. I always use NOLIST and NOWARN on PROC DATASETS.

PROC SQL Warnings

In PROC SQL, warning messages are produced if you have the same variable name twice on the select clause. This often happens when you use the SELECT * clause and join tables that have the same variables in them. You can get rid of these warnings by adding the NOWARN option to the PROC SQL statement:

```
proc sql nowarn;
```

It is important to make sure that you are getting the results that you expect from the PROC SQL code before you add the NOWARN option. It is preferable to specify the variables that you want in your output table rather than using * if at all possible, but I understand that sometimes this is impractical, so NOWARN can be a very useful tool for keeping your SAS log clean.

Type Conversion Notes

Have you seen notes like these in the SAS Log?

```
NOTE: Numeric values have been converted to character values at the
places given by:
      (Line):(Column).
      5:14
NOTE: Character values have been converted to numeric values at the
places given by:
      (Line):(Column).
      10:8
```

Most of us have. These are not warnings, but you should clear them up anyway. If you are scanning through a SAS log looking for issues, these notes will generally catch your eye and make you stop to check on them. These notes occur when a DATA step tries to use a numeric variable when it is expecting a character variable and vice versa.

If you get the numeric-to-character conversion note, this is because you have used a numeric variable or value when the statement or function was expecting a character variable. In this case, you should convert the value to character. The easiest way to do this is to use the CATS() function:

```
cats(numericVar)
```

If you get the character-to-numeric conversion note, you are trying to use a character variable or value when you should be using a numeric. In this case, you need to convert your character to a numeric value, and using the INPUT() function is the best way:

```
input(characterVar, 12.).
```

Invalid Argument Notes

An invalid argument note is not an error or a warning, but it still should be addressed because it generally means that something is wrong with your data. This is usually caused by an INPUT() function when a value is not valid for the informat, as shown in Program 6.2 and SAS Log 6.1.

Program 6.2: Invalid Argument Notes with input() Function

```
data _null_;
   x = "a";
   y = input(x, 12.);
   put y=;
run;
```

SAS Log 6.1: Invalid Argument Notes with input() Function

```
NOTE: Invalid argument to function INPUT at line 3 column 8.
y=.
x=a y=. _ERROR_=1 _N_=1
NOTE: Mathematical operations could not be performed at the
following places. The results of the operations have been set to
missing values. Each place is given by: (Number of times) at
(Line):(Column).
      1 at 1065:8
```

To suppress these notes, use the ?? modifier on the INPUT() function. When there is an invalid value, the variable is set to missing, and no invalid argument notes are put in the SAS log, as shown in Program 6.3 and SAS Log 6.2.

Program 6.3: Suppressing Invalid Argument Notes with input() Function

```
data _null_;
   x = "a";
   y = input(x, ?? 12.);
   put y=;
run;
```

SAS Log 6.2: Suppressing Invalid Argument Notes with input() Function

```
y=.
```

PROC APPEND Errors and Warnings

Error and warning messages are produced from PROC APPEND when you have variables in one of the two tables (BASE= or DATA=) but not in the other. The FORCE option is needed to force the append to occur and avoid the errors. The NOWARN option suppresses the warnings:

```
proc append base = baseTable data = newTable force nowarn;
```

You also get warnings from PROC APPEND when the lengths of the variables in the two tables don't match. If the length of a variable in the BASE= table is longer than the one in the DATA= table, the data is appended, and a warning message is put in the SAS log even if you use the FORCE option. However, if the length of a variable in the BASE= table is shorter than the one in the DATA= table, an error message is generated, and the data is not appended. To get rid of the error, you will need to change the variable length in the DATA= table.

The example in Program 6.4 and SAS Log 6.3 shows the warning messages that you can get from PROC APPEND.

Program 6.4: Warning Messages from PROC APPEND

```
data class;
   set sashelp.class;
run;

data dataTable;
   i = 1;
   name = "Charles";
   height = 70;
   output;
run;

proc append base = class data = dataTable force;
run;
```

SAS Log 6.3: Warning Messages from PROC APPEND

```
12 proc append base = class data = dataTable force;
13 run;

NOTE: Appending WORK.DATATABLE to WORK.CLASS.
WARNING: Variable i was not found on BASE file. The variable will
not be added to the BASE file.
WARNING: Variable Name has different lengths on BASE and DATA files
(BASE 8 DATA 7).
WARNING: Variable Sex was not found on DATA file.
WARNING: Variable Age was not found on DATA file.
WARNING: Variable Weight was not found on DATA file.
NOTE: FORCE is specified, so dropping/truncating will occur.
NOTE: There were 1 observations read from the data set
WORK.DATATABLE.
NOTE: 1 observations added.
NOTE: The data set WORK.CLASS has 20 observations and 5 variables.
NOTE: PROCEDURE APPEND used (Total process time):
      real time         0.01 seconds
      cpu time          0.01 seconds
```

You can get rid of some of these warnings with the NOWARN option as seen in Program 6.5 and SAS Log 6.4, but not all of them.

Program 6.5: Using the NOWARN Option with PROC APPEND

```
data class;
   set sashelp.class;
run;

data dataTable;
   i = 1;
   name = "Charles";
   height = 70;
   output;
run;

proc append base = class data = dataTable force nowarn;
run;
```

SAS Log 6.4: Using the NOWARN Option with PROC APPEND

```
12  proc append base = class data = dataTable force nowarn;
13  run;

NOTE: Appending WORK.DATATABLE to WORK.CLASS.
WARNING: Variable Name has different lengths on BASE and DATA files
(BASE 8 DATA 7).
NOTE: FORCE is specified, so dropping/truncating will occur.
NOTE: There were 1 observations read from the data set
WORK.DATATABLE.
NOTE: 1 observations added.
NOTE: The data set WORK.CLASS has 20 observations and 5 variables.
NOTE: PROCEDURE APPEND used (Total process time):
      real time         0.01 seconds
      cpu time          0.00 seconds
```

The only way to eliminate this last warning is to make sure that the DATA= table looks just like the BASE= table. I generally do this by using the following code in a DATA step to get the correct attributes for the DATA= table:

```
if (0) then set base;
```

See the "Copying Variable Attributes" section in Chapter 2 for more information about this technique.

Program 6.6 and SAS Log 6.5 show the same code without any warnings.

Program 6.6: No Warnings from PROC APPEND

```
data class;
   set sashelp.class;
run;

data dataTable;
   if (0) then
      set sashelp.class;
   call missing(of _all_);
   name = "Charles";
   height = 70;
   output;
   stop;
run;

proc append base = class data = dataTable;
run;
```

SAS Log 6.5: No Warnings from PROC APPEND

```
16  proc append base = class data = dataTable force nowarn;
17  run;

NOTE: Appending WORK.DATATABLE to WORK.CLASS.
NOTE: FORCE is specified, so dropping/truncating will occur.
NOTE: There were 1 observations read from the data set
WORK.DATATABLE.
NOTE: 1 observations added.
NOTE: The data set WORK.CLASS has 20 observations and 5 variables.
NOTE: PROCEDURE APPEND used (Total process time):
      real time         0.01 seconds
      cpu time          0.01 seconds
```

Length Warnings in a DATA Step

If you have different lengths for the same variable in a DATA step, you get warnings.

- If you have more than one LENGTH statement, the only way to eliminate the warning is to get rid of one of the LENGTH statements.
- If you set a table and then have a LENGTH statement that sets one of the variables to a different length than in the table, then you get a warning. In this case, you can move the LENGTH statement before the SET statement, and if the length set on the LENGTH statement is less than the length in the table, then you do not get a warning. However, if

the length is greater than the length of the variable in the table, then you do get a warning. To get rid of this warning, use the VARLENCHK=NOWARN SAS system option.

- If you set multiple tables that have the same variable with different lengths, and the variable with the longer length is in the earlier table, then you will get warning messages. You can get rid of these messages using the VARLENCHK=NOWARN SAS system option.

The example in Program 6.7 and SAS Log 6.6 shows a warning that you can get in a DATA step.

Program 6.7: Variable Length Warnings in a DATA Step

```
data _null_;
   length name $7;
   set sashelp.class;
run;
```

SAS Log 6.6: Variable Length Warnings in a DATA Step

```
1    data _null_;
2       length name $7;
3       set sashelp.class;
4    run;

WARNING: Multiple lengths were specified for the variable name by
         input data set(s). This can cause truncation of data.
NOTE: There were 19 observations read from the data set
SASHELP.CLASS.
NOTE: DATA statement used (Total process time):
      real time         0.00 seconds
      cpu time          0.01 seconds
```

You can get rid of this warning by setting the VARLENCHK= SAS system option temporarily to NOWARN before the DATA step, and then setting it back to WARN at the end of the DATA step. You could set it at the beginning of your program, but it helps to have the warnings turned on in case you have other length issues that need to be addressed. Program 6.8 and SAS Log 6.7 show how the VARLENCHK= option cleans up the SAS log.

Program 6.8: Using the VARLENCHK= Option to Suppress Warnings

```
options varlenchk = NOWARN;
data _null_;
   length name $7;
   set sashelp.class;
run;
options varlenchk = WARN;
```

SAS Log 6.7: Using the VARLENCHK= Option to Suppress Warnings

```
1    options varlenchk = NOWARN;
2    data _null_;
3       length name $7;
4       set sashelp.class;
5    run;
```

```
NOTE: There were 19 observations read from the data set
SASHELP.CLASS.
NOTE: DATA statement used (Total process time):
      real time       0.00 seconds
      cpu time        0.01 seconds
6   options varlenchk = NOWARN;
```

Creating Custom Error and Warning Messages

If you want to put out custom error messages and warning messages in the SAS log, you can use the DATA step PUT or PUTLOG statement:

```
putlog "ERROR: Very Bad stuff happened";
putlog "WARNING: Mildly Bad stuff happened";
```

You can also use the macro %PUT statement:

```
%put ERROR: Bad stuff happened;
%put WARNING: Mildly bad stuff happened;
```

If you use the ERROR: and WARNING: keywords in all uppercase at the beginning of the message, then in interactive SAS, these messages are color coded just as real error and warning messages produced by SAS are. Keep in mind that these error and warning messages do not stop the SAS program from processing or produce a return code like the real messages do. You need to do this yourself with an ABORT statement or by setting the &*syscc* macro variable.

> **Tip:** You can use the NOTE: keyword in the same way to get a color-coded note like a real SAS note: PUTLOG "NOTE: Something happened";

If you create custom error and warning messages, and you then do a search in the SAS log for "ERROR" or "WARNING", then you will find all the PUT statements that have the word "ERROR" or "WARNING" in them. This makes it that much harder to look for true error and warning messages.

I handle this problem by never using those two words in my code (or in my comments). Instead, I use the DATA step tricks in Program 6.9 and SAS Log 6.8, and the macro tricks in Program 6.10 and SAS Log 6.9.

Program 6.9: Custom Error and Warning Messages with a DATA Step

```
data _null_;
   putlog "WARN" "ING: Mildly Bad stuff happened";
   putlog "ERR" "OR: Very Bad stuff happened";
run;
```

SAS Log 6.8: Custom Error and Warning Messages with a DATA Step

```
1    data _null_;
2       putlog "WARN" "ING: Mildly Bad stuff happened";
3       putlog "ERR" "OR: Very Bad stuff happened";
4    run;
```

```
WARNING: Mildly Bad stuff happened
ERROR: Very Bad stuff happened
NOTE: DATA statement used (Total process time):
      real time      0.00 seconds
      cpu time       0.00 seconds
```

Program 6.10: Custom Error and Warning Messages with a Macro

```
%put WARN%str(ING:) Mildly Bad stuff happened;
%put ERR%str(OR:) Very Bad stuff happened;
```

SAS Log 6.9: Custom Error and Warning Messages with a Macro

```
1    %put WARN%str(ING:) Mildly Bad stuff happened;
WARNING: Mildly Bad stuff happened
2    %put ERR%str(OR:) Very Bad stuff happened;
ERROR: Very Bad stuff happened
```

You get the results that you want, but when you search the log you only find actual warning or error messages, not all the PUT statements.

Capturing Error and Warning Messages

If you want to save or check error or warning messages that were generated by SAS, then you can use the &SYSERRORTEXT or &SYSWARNINGTEXT automatic variables. Each of these variables holds the text of the last error or warning message that was generated. If you want to save the value, you can simply add these lines to your code:

```
%let eMsg = &syserrortext;
%let wMsg = &syswarningtext;
```

Protecting Your Password

Sometimes it is necessary to hardcode a password into your SAS program. Needless to say, this is not very secure. Even if your SAS code is protected, the password still might be displayed in your SAS log. Here are a few ways to protect a password.

Encoding a Password

You can use PROC PWENCODE to encode your password. For all your SAS passwords, this is a great way to hide them. PROC PWENCODE takes the password specified by the IN= option and generates an encoded version that it writes to the SAS Log. You can also tell the procedure to save the encoded password to a file or to the clipboard. The encoded password is the entire string that begins with "{SAS002}" (or something similar, depending on the encoding method that you have chosen). You can use this string anywhere in SAS that you would normally use a password.

The example in Program 6.11 and SAS Log 6.10 encodes the password "blue" and puts the encoded password in the SAS log.

Program 6.11: Using PROC PWENCODE to Encode a Password

```
proc pwencode in = "blue";
run;
```

SAS Log 6.10: Using PROC PWENCODE to Encode a Password

```
1   proc pwencode in = XXXXXXXXXXXX;
2   run;

{SAS002}83FBCB5C4121449726F0A94F

NOTE: PROCEDURE PWENCODE used (Total process time):
      real time         0.00 seconds
      cpu time          0.00 seconds
```

The example in Program 6.12 shows a password-protected data table being read without a password (produces an error), with the hardcoded password, and with the encoded password. You can see the results from each of these DATA steps in SAS Log 6.11.

Program 6.12: Using an Encoded Password

```
data protected (pw = blue);
   x = 1;
run;

data _null_;
   set protected;
run;

data _null_;
   set protected (pw = blue);
run;

data _null_;
   set protected (pw = '{SAS002}83FBCB5C4121449726F0A94F');
run;
```

SAS Log 6.11: Using an Encoded Password

```
1   data protected (pw = XXXX);
2       x = 1;
3   run;

NOTE: The data set WORK.PROTECTED has 1 observations and 1
variables.
NOTE: DATA statement used (Total process time):
      real time         0.01 seconds
      cpu time          0.01 seconds

4
5   data _null_;
6       set protected;
ERROR: Invalid or missing READ password on member
WORK.PROTECTED.DATA.
7   run;

NOTE: The SAS System stopped processing this step because of errors.
NOTE: DATA statement used (Total process time):
   real time         1.49 seconds
   cpu time          0.07 seconds
```

```
8
9    data _null_;
10      set protected (pw = XXXX);
11   run;

NOTE: There were 1 observations read from the data set
WORK.PROTECTED.
NOTE: DATA statement used (Total process time):
      real time       0.00 seconds
      cpu time        0.00 seconds

12
13   data _null_;
14      set protected (pw = XXXXXXXXXXXXXXXXXXXXXXXXXXXXXXXXXXXX);
15   run;

NOTE: There were 1 observations read from the data set
WORK.PROTECTED.
NOTE: DATA statement used (Total process time):
      real time       0.00 seconds
      cpu time        0.00 seconds
```

Storing a Password

Another way to hide a password is to put it in a file that is protected using the operating system permissions so that the person running the program must have Read permission to the file. The password can then be read from the file in the SAS program and used as needed. For double security, you can save an encoded password in a file.

The first step is to store the password in a file, and then protect the file using the operating system's protections. Figure 6.1 shows the password file, pwd.txt, in Notepad.

Figure 6.1: Storing the Password in a File Called pwd.txt

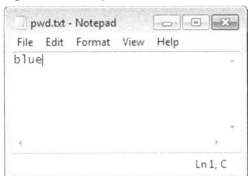

The code in Program 6.13 can now read this file and create a macro variable that contains the password. This variable can then be used to read the protected table that is created in the first DATA step. The output from this program is shown in SAS Log 6.12.

Program 6.13: Using a Stored Password

```
data protected (pw = blue);
   x = 1;
run;

data _null_;
   infile "c:\temp\pwd.txt" truncover;
   input line $50.;
   call symputx("pwd", line);
run;

data _null_;
   set protected (pw = &pwd);
run;
```

SAS Log 6.12: Using a Stored Password

```
1    data protected (pw = XXXX);
2       x = 1;
3    run;

NOTE: The data set WORK.PROTECTED has 1 observations and 1
variables.
NOTE: DATA statement used (Total process time):
      real time       0.01 seconds
      cpu time        0.01 seconds


4
5    data _null_;
6       infile "c:\temp\pwd.txt" truncover;
7       input line $50.;
8       call symputx("pwd", line);
9    run;

NOTE: The infile "c:\temp\pwd.txt" is:
      Filename=c:\temp\pwd.txt,
      RECFM=V,LRECL=32767,File Size (bytes)=6,
      Last Modified=14Apr2017:13:36:21,
      Create Time=14Apr2017:12:43:36

NOTE: 1 record was read from the infile "c:\temp\pwd.txt".
      The minimum record length was 4.
      The maximum record length was 4.
NOTE: DATA statement used (Total process time):
      real time       0.00 seconds
      cpu time        0.01 seconds


10
11   data _null_;
12      set protected (pw = &pwd);
13   run;

NOTE: There were 1 observations read from the data set
WORK.PROTECTED.
```

```
NOTE: DATA statement used (Total process time):
      real time         0.00 seconds
      cpu time          0.00 seconds
```

Hiding a Password

To be extra safe, you need to make sure that the password doesn't get printed out in the SAS log. Because users could set all the logging options on and then run your code, it is possible that they could see a password in your code, particularly if you use macros.

The code in Program 6.14 is a macro that reads the password file and creates a macro variable with the password. If the MPRINT and SYMBOLGEN options are turned on, the password is printed to the log. The second DATA step shows that when you hardcode the password, it can show up in the SAS log, as you can see in SAS Log 6.13.

Program 6.14: Passwords in the SAS Log

```
options mprint symbolgen;

%macro testPassword;

   data protected (pw = blue);
      x = 1;
   run;

   data _null_;
      infile "c:\temp\pwd.txt" truncover;
      input line $50.;
      call symputx("pwd", line);
   run;

   data _null_;
      set protected (pw = &pwd);
   run;

   data _null_;
      set protected (pw = blue);
   run;

%mend testPassword;

%testPassword;
```

SAS Log 6.13: Passwords in the SAS Log

```
MPRINT(TESTPASSWORD):   data protected (pw = blue);
MPRINT(TESTPASSWORD):   x = 1;
MPRINT(TESTPASSWORD):   run;

NOTE: The data set WORK.PROTECTED has 1 observations and 1
variables.
NOTE: DATA statement used (Total process time):
      real time         0.01 seconds
      cpu time          0.01 seconds
```

```
MPRINT(TESTPASSWORD):  data _null_;
MPRINT(TESTPASSWORD):  infile "c:\temp\pwd.txt" truncover;
MPRINT(TESTPASSWORD):  input line $50.;
MPRINT(TESTPASSWORD):  call symputx("pwd", line);
MPRINT(TESTPASSWORD):  run;

NOTE: The infile "c:\temp\pwd.txt" is:
      Filename=c:\temp\pwd.txt,
      RECFM=V,LRECL=32767,File Size (bytes)=6,
      Last Modified=14Apr2017:13:36:21,
      Create Time=14Apr2017:12:43:36

NOTE: 1 record was read from the infile "c:\temp\pwd.txt".
      The minimum record length was 4.
      The maximum record length was 4.
NOTE: DATA statement used (Total process time):
      real time         0.00 seconds
      cpu time          0.01 seconds

MPRINT(TESTPASSWORD):  data _null_;
SYMBOLGEN: Macro variable PWD resolves to blue
MPRINT(TESTPASSWORD):  set protected (pw = blue);
MPRINT(TESTPASSWORD):  run;

NOTE: There were 1 observations read from the data set
WORK.PROTECTED.
NOTE: DATA statement used (Total process time):
      real time         0.00 seconds
      cpu time          0.00 seconds

MPRINT(TESTPASSWORD):  data _null_;
MPRINT(TESTPASSWORD):  set protected (pw = blue);
MPRINT(TESTPASSWORD):  run;

NOTE: There were 1 observations read from the data set
WORK.PROTECTED.
NOTE: DATA statement used (Total process time):
      real time         0.00 seconds
      cpu time          0.00 seconds
```

If this DATA step code were run from outside a macro, then the passwords would be masked with XXXX, but the MPRINT option does not mask the passwords for you. To make sure that the passwords are not printed in the SAS log, you need to turn off the MPRINT and SYMBOLGEN options at the beginning of your macro. In the code in Program 6.15, I save the original settings of the options before turning them off. I then reset them at the end of the macro. The output in SAS Log 6.14 shows that the password is no longer displayed in the SAS log.

Program 6.15: Turning Off MPRINT and SYMBOLGEN Options

```
options mprint symbolgen;

%macro testPassword;

   %let mprint = %sysfunc(getoption(mprint));
   %let symbolgen = %sysfunc(getoption(symbolgen));

   options nomprint nosymbolgen;

   data protected (pw = blue);
      x = 1;
   run;

   data _null_;
      infile "c:\temp\pwd.txt" truncover;
      input line $50.;
      call symputx("pwd", line);
   run;

   data _null_;
      set protected (pw = &pwd);
   run;

   data _null_;
      set protected (pw = blue);
   run;

   options &mprint &symbolgen;

%mend testPassword;

%testPassword;
```

SAS Log 6.14: Turning Off MPRINT and SYMBOLGEN Options

```
MPRINT(TESTPASSWORD):  options nomprint nosymbolgen

NOTE: The data set WORK.PROTECTED has 1 observations and 1
variables.
NOTE: DATA statement used (Total process time):
      real time        0.01 seconds
      cpu time         0.00 seconds

NOTE: The infile "c:\temp\pwd.txt" is:
      Filename=c:\temp\pwd.txt,
      RECFM=V,LRECL=32767,File Size (bytes)=6,
      Last Modified=14Apr2017:13:36:21,
      Create Time=14Apr2017:12:43:36

NOTE: 1 record was read from the infile "c:\temp\pwd.txt".
      The minimum record length was 4.
      The maximum record length was 4.
NOTE: DATA statement used (Total process time):
      real time        0.00 seconds
      cpu time         0.01 seconds
```

```
NOTE: There were 1 observations read from the data set
WORK.PROTECTED.
NOTE: DATA statement used (Total process time):
      real time         0.00 seconds
      cpu time          0.00 seconds

NOTE: There were 1 observations read from the data set
WORK.PROTECTED.
NOTE: DATA statement used (Total process time):
      real time         0.00 seconds
      cpu time          0.00 seconds
```

Using the PUT, PUTLOG, and %PUT statements

The PUT and PUTLOG statements are very useful in a DATA step, as is the %PUT macro statement. These are life-savers for debugging to display values in the SAS log that help trace what is happening. In addition, if you are creating an output file or report, the PUT statement enables you to lay out the contents with a lot of precision that is not as easy with PROC PRINT or PROC REPORT.

Tip: Even if you are not debugging your program with PUT statements, you should always check the log to ensure that your code ran correctly. It is easy to skip this step and not realize that the code had an error early on that you didn't notice, or a programming issue that caused a table to have the wrong contents.

Debugging with PUTLOG

As you'll see from most of the examples in this book, I use PUTLOG anytime I want to put out some information in the SAS log. You can also use the PUT statement to do this. However, if you use a PUT statement with debugging information after a FILE statement in the DATA step to write to an external file or to the output window, the debugging information is written to the file instead of to the log. To get around this problem, you can specify FILE LOG before the PUT statement, and then set the FILE statement back to the real file afterwards. This is a lot of work! Instead, you can just use PUTLOG to put the information to the log. Then it doesn't matter whether there is a FILE statement or not. My suggestion is to get in the habit of using PUTLOG when you want to write to the SAS log.

Using the VAR= Syntax

Writing the value of a variable to the log is a very useful debugging tool. You can use the following syntax to write the value of a variable to the log:

```
putlog var;
```

If you are writing out multiple values, it can be confusing to determine which value is which. So you can use the VAR= syntax instead. This syntax automatically puts the variable name, an equal sign, and the value in the log:

```
putlog var=;
```

The SAS log displays the following:

```
var=ABC
```

You can also use the VAR= syntax when writing to a file with a PUT statement.

Using Shortcuts

The program data vector (PDV) contains all the variables that are defined in the DATA step. If you want to see the values of all the variables in the PDV, you can use the _ALL_ shortcut with PUT or PUTLOG:

```
putlog _ALL_;
```

This causes all the variables and values to be displayed, including the automatic variables such as the FIRST., and LAST., and _N_. The output looks something like this:

```
Name=Alice Sex=F Age=13 Height=56.5 Weight=84 FIRST.Sex=1
LAST.Sex=0 _ERROR_=0 _N_=1
```

In addition to _ALL_, you can also use _CHARACTER_ or _NUMERIC_ to see all the character or numeric variables. When you are debugging a program, these shortcuts come in handy.

Getting Rid of Unneeded Spaces

The PUT and PUTLOG statements automatically add a space after a variable value. The following code illustrates this:

```
data _null_;
   var = "123";
   put "***" var "***";
run;
```

This code creates the following line:

```
***123 ***
```

To remove the space after the "3", you need to tell the PUT statement to back up by 1 space before adding the three *s. To do this, use the +(-1) syntax:

```
data _null_;
   var = "123";
   put "***" var +(-1) "***";
run;
```

Then you get the following line:

```
***123***
```

The + tells the PUT statement to move the pointer, and the value after the + tells the PUT statement how many spaces to move. In this case, we want to move backward one space, so we use -1. The parentheses are needed to avoid syntax errors. If you want to move the pointer forward 3 spaces, you would specify +3. The parentheses are not needed for positive numbers.

Displaying a Macro Value in the SAS Log

If you want to display the value of a macro variable in the SAS log, you use the %PUT statement:

```
%put &test;
```

If you want to make it more meaningful, you can add the name of the macro variable as well:

```
%put test = &test;
```

Alternatively, you can use the &= syntax to put the name of the variable and its value in the log (similar to PUTLOG VAR= in a DATA step):

```
%put &=test;
```

This syntax produces the following:

```
TEST=abc
```

Using Macro Shortcuts

You might want to know what macro variables you have created and what they are set to—something like the results of a PUTLOG _ALL_; in a DATA step. The following examples show how you can do this.

To see all your macro variables, use the following code:

```
%put _all_;
```

To see all your global macro variables, use the following code:

```
%put _global_;
```

To see all your local macro variables (variables that are only available in the execution of the current macro), use the following code:

```
%put _local_;
```

To see all the automatic macro variables (the ones that SAS creates for you), use the following code:

```
%put _automatic_;
```

Chapter 7: Advanced Tasks

Introduction

This chapter includes examples of more advanced tasks that you can do with SAS. These examples are useful as demonstrations of what you can make SAS software do for you.

Sending Email

I often send emails from a SAS program, particularly to report success, warnings, and errors for automated production programs. The only caveat when you use SAS to send emails is that you must be using the Simple Mail Transfer Protocol (SMTP) email interface. You need to have the EMAILHOST= and EMAILSYS= options set in your SAS session to send emails, as follows:

```
options emailhost = <SMTP.server> emailsys = SMTP;
```

These options might have been set during the installation of SAS; you can check the value by submitting this code:

```
proc options option = emailhost;
run;
```

Note that the email addresses in the following examples are all invalid, so this code does not actually work. Substitute real email addresses if you want to run the code.

A Single Message

The easiest way to send email is as a single message. You can use a FILENAME statement with the email access method and specify the email address, subject, attachments, and other options on the FILENAME statement. Then you use a DATA step to write the text of the email to the filename, and that's all there is to it. Program 7.1 and SAS Log 7.1 show how to send an email. Figure 7.1 shows what the email looks like in Microsoft Outlook.

Program 7.1: Sending a Single Email

```
filename em email to = "user@company.com"
                  subject = "SAS Email Test";
data _null_;
   set sashelp.class;
   where age gt 14;
   file em;
   if (_n_ eq 1) then
      put "Students older than 14:";
   put @3 name @12 age;
run;
filename em;
```

SAS Log 7.1: Sending a Single Email

```
NOTE: The file EM is:
      E-Mail Access Device

Message sent
      To:          "user@company.com"
      Cc:
      Bcc:
      Subject:   SAS Email Test
      Attachments:
NOTE: 6 records were written to the file EM.
      The minimum record length was 13.
      The maximum record length was 23.
NOTE: There were 5 observations read from the data set
      SASHELP.CLASS.
      WHERE age>14;
```

Figure 7.1: Sending a Single Email

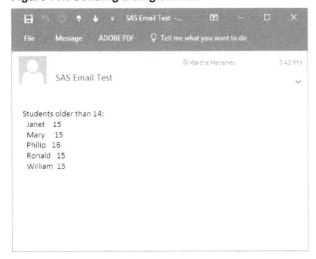

Multiple Messages

To send different emails to different people on the basis of the contents of a SAS data table, you can use email directives. These directives are keywords that you put into the body of the email to specify information such as the email address and subject. You still use the FILENAME statement with the email access method, but you don't need to specify any options on the FILENAME statement. Here are some basic directives:

!EM_TO!
> specifies the email addresses to send to.

!EM_FROM!
> specifies the email address that the email should come from (for example, noreply@sas.com).

!EM_SUBJECT!
> specifies the subject.

!EM_ATTACH!
> specifies a file to be attached to the email.

When you are using the email access engine with email directives, the DATA step automatically sends the email at the end of each iteration. If you don't want this automatic process to happen, here are some additional directives that you need:

!EM_SEND!
> indicates the end of an email.

!EM_NEWMSG!
> indicates that a new email is beginning (clears all existing directive values).

!EM_ABORT!
> is used at the end of the DATA step to stop the automatic send email action. When you want to use !EM_SEND! to send the email when you are ready, use !EM_ABORT! at the end of the DATA step.

The following example illustrates sending an email to the class in the sashelp.class table. The girls in the class get one email, and the boys get a different one. Program 7.2 handles the first step in the process by creating a data table with the two emails to send, including the email list, the subject, the FROM email address, and the email text.

Program 7.2: Creating the Data Table to Send Multiple Emails

```
proc sort data = sashelp.class
           out = class;
   by sex;
run;

data class (keep = sex emailList emailText subject from);
   set class;
   by sex;
   length emailText $100;
   length emailList $500;
   retain emailList;
   if (first.sex) then
      emailList = "";
   emailList = catx(" ", emailList,
               cats('"', name, "@OurSchool.edu", '"'));  ❶
   if (last.sex) then
   do;

      subject = "Test!";        ❷
      from = "Your.Teacher@OurSchool.edu";   ❸

      if (sex eq "F") then      ❹
         emailText = "Ladies,";
      else
         emailText = "Gentlemen,";
      output;
    ❺
      emailText = "Next Monday we will have a test on the topics";
      output;
      emailText = " that we will be studying this week.";
      output;
      emailText = "Prepare carefully!";
      output;
      emailText = "Sincerely,";
      output;
      emailText = "  Your Teacher";
      output;

   end; /* if - last record of this value of sex */

run;
```

❶ Create a list of the email addresses to send the email to. Each email address should be in quotation marks, and they should be delimited with spaces.

❷ Set the email's subject.

❸ Set the FROM email address.

❹ Set the first line of text on the email on the basis of the sex.

❺ Create the text of the email.

The second step in the process is in Program 7.3, which uses the table created in Program 7.2 to send the two emails. SAS Log 7.2 shows the output after running the DATA step.

Program 7.3: Sending Multiple Emails

```
filename em email;   ❶

data _null_;
   set class end = eof;
   by sex;
   file em;

   if (first.sex) then     ❷
   do;
      put "!EM_SUBJECT! " subject;
      put "!EM_TO! " emailList;
      put "!EM_FROM! " from;
   end;

   put emailText $char100.;   ❸

   if (last.sex) then   ❹
   do;
      put "!EM_SEND!";
      put "!EM_NEWMSG!";
   end; /* if - last record for this group */

   put "!EM_ABORT!";   ❺

run;

filename em;
```

❶ Use a FILENAME statement with the email device. No other options are necessary.
❷ For the first record of the sex, start the email with the subject, TO email addresses, and FROM email address.
❸ Use PUT to write each line of email text.
❹ At the last record of this sex, issue the SEND directive and then clear all the directives to start a new email.
❺ Issue the ABORT directive, since we don't want to send an email until after all the email text has been written to the email.

SAS Log 7.2: Sending Multiple Emails

```
NOTE: The file EM is:
      E-Mail Access Device

Message sent
      To:     "Alice@OurSchool.edu" "Barbara@OurSchool.edu"
"Carol@OurSchool.edu" "Jane@OurSchool.edu" "Janet@OurSchool.edu"
"Joyce@OurSchool.edu" "Judy@OurSchool.edu" "Louise@OurSchool.edu"
"Mary@OurSchool.edu"
   Cc:
   Bcc:
   Subject:   Test!
   Attachments:
```

```
Message sent
     To:      "Alfred@OurSchool.edu" "Henry@OurSchool.edu"
"James@OurSchool.edu" "Jeffrey@OurSchool.edu" "John@OurSchool.edu"
"Philip@OurSchool.edu" "Robert@OurSchool.edu" "Ronald@OurSchool.edu"
"Thomas@OurSchool.edu" "William@OurSchool.edu"
     Cc:
     Bcc:
     Subject:    Test!
     Attachments:
NOTE: 34 records were written to the file EM.
```

Tip: These techniques for sending emails also work to send text messages. Just replace the email address with a mobile phone number (with no spaces, hyphens or parentheses) and add the appropriate Short Message Service (SMS) gateway following an @. Each carrier has a different SMS gateway; check online to find the correct one. Here is an example:

9195558000@txt.att.net

(NOTE: This is NOT a real phone number.)

Running Code in Parallel

When you are running on a machine that has multiple processors, it can be to your advantage to run pieces of your code in parallel on the different processors. The different pieces of code must be independent of each other; that is, they can't rely on data produced by code that is running in parallel.

The tool to do this is MPCONNECT. It uses options and syntax from SAS/CONNECT, but runs the code on the same server, rather than running on a different server.

Using MPCONNECT

The best way to explain how to use MPCONNECT is to show you a simple example.

For this example, I read two tables from sashelp: the flags and demographics tables. I need to do some processing of each table, and after that is complete, I want to merge the new tables. The processing of each table can be done in parallel, and after both of those processes are complete, the tables can be merged. I've added a macro variable that is created in the primary (local) session and is copied to the remote sessions using the %SYSLPUT statement. I've also added a macro variable from one of the remote sessions that is copied back to the primary session using the %SYSRPUT statement, just to show how it is done. The tables and processing in this example are far too small to warrant using MPCONNECT, but if the tables were much larger and the processing much more involved, this would be a good way to improve performance.

The code for this basic MPCONNECT example is in Program 7.4, with some notes about the code afterwards.

Program 7.4: Basic MPCONNECT Program

```
%let firstLetter = C;

options sascmd = "!sascmd";   ❶
signon flag_ses;

%syslput first = &firstLetter / remote = flag_ses;   ❷

rsubmit flag_ses wait = no
                 inheritlib = (work = pwork);   ❸

   %let nobs = 0;                    ❹
   data pwork.flags (keep = country flagCatalog);
      set sashelp.flags end = eof;
      where substr(title, 1, 1) eq "&first";
      country = propcase(title);
      length flagCatalog $35;
      flagCatalog = catx(".", upcase(location), file, "IMAGE");
      if (eof) then
         call symputx("nobs", _n_);
   run;

   proc sort data = pwork.flags;
      by country;
   run;

   %sysrput flags_nobs = &nobs;   ❺

endrsubmit;

signon demo_ses;   ❻

%syslput fl = &firstLetter / remote = demo_ses;

rsubmit demo_ses wait = no
                 inheritlib = (work = pwork);

   data pwork.demographics (keep = country region);
      set sashelp.demographics;
      where substr(name, 1, 1) eq "&fl";
      country = propcase(name);
   run;

   proc sort data = pwork.demographics;
      by country;
   run;

endrsubmit;

waitfor _all_     ❼
        flag_ses
        demo_ses;

signoff _all_;   ❽

%put &=flags_nobs;
```

```
data regional_flags (keep = region country flagCatalog);  ❾
   merge flags demographics;
   by country;
run;
```

❶ Specify the command to start a new SAS session, and then the SIGNON statement to create the first SAS session and sign on to it. Note that a session name (flag_ses in this example) cannot be more than 8 characters.

❷ Create a macro variable (called first) in the remote SAS session based on the value of a macro variable from the local SAS session (&firstLetter).

❸ Submit the code between the RSUBMIT and the ENDRSUBMIT statements to the remote SAS session. Use the WAIT=NO option to keep processing the code in the local session, and the INHERITLIB= option to point to libraries in the local SAS session. In this example, the local session's work library can be accessed with the libref pwork, which enables the remote session to read and write to the local session's work library.

❹ Process the flags table: Subset by the &first macro variable, and save the new table in the pwork library (the local session's work library).

❺ Use %SYSRPUT to create a macro variable in the local SAS session (&flag_nobs) with the value from the remote session (&nobs).

❻ Create the second SAS session and sign on to it. Proceed with creating the macro variable in the remote session and submitting the code to process the demographics table to the remote SAS session.

❼ Wait for the remote sessions to finish before continuing.

❽ Sign off all the remote sessions.

❾ Merge the processed flags and demographics tables together.

The log from the MPCONNECT program is in SAS Log 7.3. As you can see, logs from MPCONNECT are quite long and hard to follow because of the multiple SAS sessions.

SAS Log 7.3: Basic MPCONNECT Program

```
1    %let firstLetter = C;
2
3    options sascmd = "!sascmd";
4
5    signon flag_ses;
NOTE: Remote signon to FLAG_SES commencing (SAS Release
9.04.01M0P061913).
NOTE: Unable to open SASUSER.PROFILE. WORK.PROFILE will be opened
instead.
NOTE: All profile changes will be lost at the end of the session.
NOTE: Copyright (c) 2002-2012 by SAS Institute Inc., Cary, NC, USA.
NOTE: SAS (r) Proprietary Software 9.4 (TS1M0)
NOTE: This session is executing on the X64_7PRO platform.

NOTE: SAS initialization used:
      real time       0.69 seconds
      cpu time        0.70 seconds

NOTE: Remote signon to FLAG_SES complete.
```

```
6
7   %syslput first = &firstLetter / remote = flag_ses;
8
9   rsubmit flag_ses wait = no
10                  inheritlib = (work = pwork);
NOTE: Background remote submit to FLAG_SES in progress.
11
12  signon demo_ses;
NOTE: Remote signon to DEMO_SES commencing (SAS Release
9.04.01M0P061913).
NOTE: Unable to open SASUSER.PROFILE. WORK.PROFILE will be opened
instead.
NOTE: All profile changes will be lost at the end of the session.
NOTE: Copyright (c) 2002-2012 by SAS Institute Inc., Cary, NC, USA.
NOTE: SAS (r) Proprietary Software 9.4 (TS1M0)

NOTE: SAS initialization used:
      real time         0.65 seconds
      cpu time          0.65 seconds

NOTE: Remote signon to DEMO_SES complete.
13
14  %syslput fl = &firstLetter / remote = demo_ses;
15
16  rsubmit demo_ses wait = no
17                  inheritlib = (work = pwork);
NOTE: Background remote submit to DEMO_SES in progress.
18
19  waitfor _all_
20          flag_ses
21          demo_ses;
22
23  signoff _all_;
NOTE: Remote submit to FLAG_SES commencing.
1       %let nobs = 0;
2       data pwork.flags (keep = country flagCatalog);
3           set sashelp.flags end = eof;
4           where substr(title, 1, 1) eq "&first";
5           country = propcase(title);
6           length flagCatalog $35;
7           flagCatalog = catx(".", upcase(location), file, "IMAGE");
8           if (eof) then
9               call symputx("nobs", _n_);
10      run;

NOTE: There were 19 observations read from the data set
SASHELP.FLAGS.
   WHERE SUBSTR(title, 1, 1)='C';
NOTE: The data set PWORK.FLAGS has 19 observations and 2 variables.
NOTE: DATA statement used (Total process time):
      real time         0.03 seconds
      cpu time          0.01 seconds
```

```
11
12     proc sort data = pwork.flags;
13        by country;
14     run;

NOTE: There were 19 observations read from the data set PWORK.FLAGS.
NOTE: The data set PWORK.FLAGS has 19 observations and 2 variables.
NOTE: PROCEDURE SORT used (Total process time):
      real time        0.01 seconds
      cpu time         0.00 seconds


15
16      %sysrput flags_nobs = &nobs;
NOTE: Remote submit to FLAG_SES complete.
NOTE: Remote signoff from FLAG_SES commencing.
NOTE: Remote submit to DEMO_SES commencing.
1        data pwork.demographics (keep = country region);
2           set sashelp.demographics;
3           where substr(name, 1, 1) eq "&fl";
4           country = propcase(name);
5        run;

NOTE: There were 18 observations read from the data set
SASHELP.DEMOGRAPHICS.
      WHERE SUBSTR(name, 1, 1)='C';
NOTE: The data set PWORK.DEMOGRAPHICS has 18 observations and 2
variables.
NOTE: DATA statement used (Total process time):
      real time        0.03 seconds
      cpu time         0.03 seconds


6
7        proc sort data = pwork.demographics;
8           by country;
9        run;

NOTE: There were 18 observations read from the data set
PWORK.DEMOGRAPHICS.
NOTE: The data set PWORK.DEMOGRAPHICS has 18 observations and 2
variables.
NOTE: PROCEDURE SORT used (Total process time):
      real time        0.01 seconds
      cpu time         0.00 seconds


NOTE: Remote submit to DEMO_SES complete.
NOTE: Remote signoff from DEMO_SES commencing.
NOTE: SAS Institute Inc., SAS Campus Drive, Cary, NC USA 27513-2414
NOTE: The SAS System used:
      real time        2.85 seconds
      cpu time         0.76 seconds


NOTE: SAS Institute Inc., SAS Campus Drive, Cary, NC USA 27513-2414
NOTE: The SAS System used:
      real time        1.09 seconds
      cpu time         0.73 seconds


NOTE: Remote signoff from FLAG_SES complete.
```

```
NOTE: Remote signoff from DEMO_SES complete.
24
25  %put &=flags_nobs;
FLAGS_NOBS=19
26
27  data regional_flags (keep = region country flagCatalog);
28     merge flags demographics;
29     by country;
30  run;

NOTE: There were 19 observations read from the data set WORK.FLAGS.
NOTE: There were 18 observations read from the data set
WORK.DEMOGRAPHICS.
NOTE: The data set WORK.REGIONAL_FLAGS has 23 observations and 3
variables.
NOTE: DATA statement used (Total process time):
      real time         0.02 seconds
      cpu time          0.00 seconds
```

Tip: Notice the PROPCASE() function that I used in this example. It is like UPCASE() and LOWCASE(), except it converts all the words in a string to proper case. So a string such as "martha MESSINEO" becomes "Martha Messineo".

Striping Data

Data striping is a great use of MPCONNECT when you are working with really large data tables (those with millions of rows). This method enables you to divide the table into chunks (or stripes), process the chunks in parallel, and then bring all the chunks back together. I have seen big performance improvements using this method.

The following is an example of data striping. Program 7.5 is run first to create a data table with five million records to use in the example.

Program 7.5: Creating a Large Data Table for Data Striping

```
libname big "C:\temp";

data big.bigData;
   do numValue = 1 to 5000000;
      charValue = cats("CHAR", numValue);
      output;
   end;
run;
```

The code in Program 7.6 determines how many processors you have available and makes that many data stripes with the same number of records in each stripe (or as close as possible). I've added notes after the code to help describe the process, but I am not including the SAS log because it is just too long. This code can be run from any SAS session if you want to try it. Just run the code in Program 7.5 first and change the LIBNAME at the top to point to an appropriate place on your system.

Program 7.6: Using MPCONNECT for Data Striping

```
%macro stripe;

   %let numobs = 0;                    ❶
   %let dsid = %sysfunc(open(big.bigData));
   %if (&dsid ne 0) %then
   %do;
      %let numobs = %sysfunc(attrn(&dsid, NLOBS));
      %let dsid = %sysfunc(close(&dsid));
   %end;

   %if (&numobs eq 0) %then
      %return;

   %let numCPU = &sysncpu;      ❷

   %let numObsPerSession = %eval(&numobs / &numCPU);

   options sascmd="!sascmd";

   %let firstobs = 1;          ❸
   %let obs = &numObsPerSession;

   %do i = 1 %to &numCPU;    ❹

      signon session&i;

      %syslput session = &i / remote = session&i; ❺
      %syslput firstobs = %str(firstobs = &firstobs) /
               remote = session&i;

      %if (&i eq &numCPU) %then              ❻
         %syslput obs = %str() / remote = session&i;
      %else
         %syslput obs = %str(obs=&obs) / remote = session&i;

      rsubmit session&i wait=no
                        inheritlib = (work = pwork
                                      big = big);   ¼

         data pwork.subset&session;          ½
            set big.bigData (&firstobs &obs);
            newVar = int(ranuni(_n_) * 1000);
         run;

      endrsubmit;

      %let firstobs = %eval(&firstobs + &numObsPerSession);   ❾
      %let obs = %eval(&obs + &numObsPerSession);

   %end; /* do i - loop through processors/sessions */
```

```
      waitfor _all_
      %do i = 1 %to &numCPU;
         session&i
      %end;
      ;

   signoff _all_;

   %do i = 2 %to &numCPU;         ⑩
      proc append base = subset1
                  data = subset&i;
      run;
   %end; /* do i - loop through processors/sessions */

%mend stripe;
%stripe;
```

❶ Get the total number of records in the table, and stop processing if there are no records in the table.

❷ Get the number of processors on this machine; a SAS session is created on each processor. Adjust this number if you don't want to use all the processors. Then calculate the number of records to process in each SAS session.

❸ Set &firstobs to 1 and &obs to the number of records in each data stripe for the first data stripe.

❹ Loop through the processors and create a remote session on each one using &i in the session name to distinguish between them.

❺ Send the session number into the remote session. This number is used to name the output table that is created. Create the &firstobs macro variable that is used to subset the data into the appropriate stripe.

❻ Create the &obs macro variable to use for subsetting the data into the appropriate number of records. If this is the last data stripe, there is no need to specify the number of records. Just read from the &firstobs value to the end of the table.

❼ Submit the code from the RSUBMIT statement to the ENDRSUBMIT statement to the remote session. Link the pwork libref to the local session's work library and the big libref to the big libref in the local session.

❽ Subset the big data into the data stripe using the FIRSTOBS= and OBS= options on the SET statement, and process the data. In this example, I am adding a new variable to the table.

❾ Calculate the next &firstobs value (the current &firstobs value plus the number of records), and the next &obs value (the current &obs value plus the number of records). These are used to create the next data stripe.

⑩ Loop through all the sessions and append the data stripes together to re-create the big table.

Simulating Recursion

Recursion is sometimes the best method to use when you are working through structures that have an unknown size (for example, a directory structure with multiple subdirectories and files at many levels). Recursion allows a piece of code to call itself as many times as necessary. Recursion is not built into SAS because the scope of variables is not as limited as it is in other languages. However, you can simulate recursion with a macro by keeping track of how deep you are in macro calls and using macro arrays to hold the values that you need at the different levels. A macro array is simply a set of macro variables that have a number on the end of the name so that you can iterate over the set, as shown in this example:

```
%do i = 1 %to 5;
    %let var&i = &i;
%end;
```

This code creates the macro variables: &var1, &var2, &var3, &var4, and &var5, which you can then iterate over as follows:

```
%do i = 1 %to 5;
    %put &&var&i;
%end;
```

To simulate recursion, you can use a counter to keep track of how deep you are in the recursion, and macro arrays to keep track of any values that you need to remember. You can call the macro from within itself as many times as you like.

The example in Program 7.7 traverses a Windows directory structure and saves all the filenames that are found. A macro variable called &level keeps track of the level of the macro that is currently running. Every time the macro is called, the &level variable is incremented by 1, and every time processing returns to the calling macro, the &level variable is decremented by 1. There are macro arrays to hold the directory ID variable, the looping variable, and the path that this iteration of the macro started with. The macro keeps a counter (&fCount) of all the files that are found, and each file's path is stored in another macro array called &&file&fCount.

Program 7.7: Using Recursion to Traverse a Directory Tree

```
/* get the correct type of folder delimiter for Windows or Unix */
%let s = %sysfunc(ifc(&sysscp eq WIN, %str(\), %str(/)));

/* count of number of files */
%let fCount = 0;

/* starting level */
%let level = 0;

%macro traverseTree(startPath);

    %let level = %eval(&level + 1);      ❶

    %let startPath&level = &startPath;   ❷

    %let fname = m&level;   ❸
    %let rc = %sysfunc(filename(fname, %bquote(&&startPath&level)));
    %let did&level = %sysfunc(dopen(&fname));

    %if (&&did&level le 0) %then   ❹
    %do;

        %let fCount = %eval(&fCount + 1);
        %global file&fCount;
        %let file&fCount = &&startPath&level;
```

```
          %let fname = m&level;
          %let rc = %sysfunc(filename(fname));

          %return;   ❺

      %end; /* if - not a directory */

      %do i&level = 1 %to %sysfunc(dnum(&&did&level));   ❻

          %let nm = %bquote(%sysfunc(dread(&&did&level, &&i&level)));   ❼

          %traverseTree(%bquote(&&startPath&level&s&nm));   ❽

          %let level = %eval(&level - 1);   ❾

      %end; /* do i&level - loop through directory members */

      %let did&level = %sysfunc(dclose(&&did&level));

      %let fname = m&level;
      %let rc = %sysfunc(filename(fname));

%mend traverseTree;
%traverseTree(c:\temp);
```

❶ Increment the level counter for the current iteration.

❷ Save the path at which this iteration is starting.

❸ Attempt to open the path as a directory.

❹ If it isn't possible to open the path as a directory, then the path is a file, which means that this is the end of a branch. So save the file and go back to the previous iteration and get the next member.

❺ Stop the current iteration of this macro and return to the calling program (which might be a previous iteration of the macro).

❻ If the path is a directory, then loop through the members.

❼ Get the name of the member and create the full path.

❽ Recurse: Call the macro again, using the full path and the member name as the starting place.

❾ When returning from the recursive macro call, decrement &level by 1 to get back to the current iteration number.

After the recursion is complete, you can loop through the list of files and process them as needed. Program 7.8 shows how to list all the files that were found, and SAS Log 7.4 shows the files that I found when I ran these programs.

Program 7.8: Displaying the Files Found

```
%macro listFiles;
    %do i = 1 %to &fCount;
        %put &i &&file&i;
    %end;
%mend;
%listFiles;
```

SAS Log 7.4: Displaying the Files Found

```
1 c:\temp\ABC.txt
2 c:\temp\class\Alfred\Test.txt
3 c:\temp\class\James\Test.txt
4 c:\temp\class\Robert\Test.txt
5 c:\temp\class\William\Test\Test.txt
6 c:\temp\Test.txt
```

Reading Metadata

The SAS Metadata Repository (formerly called the *SAS Open Metadata Repository*, or OMR) is used by lots of SAS products to store metadata (data about the system). You can access the metadata from a SAS program to retrieve information about libraries, tables, users, and other objects. Why do you care? Because, as an example, you can then use this information to programmatically generate LIBNAME statements and to create tables with the correct columns and attributes. I have also used the user metadata to verify that users are in the correct metadata groups and have the right permissions. There are many other reasons to read metadata, but getting data table and user information are the most common from within a SAS program.

From a SAS program, you can use PROC METADATA, which requires you to write XML to define the data to retrieve, or you can use DATA step functions, which enable you to programmatically traverse through the metadata. I prefer using the DATA step functions, because I think the resulting program is easier to write and debug. The rest of this section demonstrates this method.

First, you need a basic understanding of SAS metadata. Metadata is stored in a web-like structure with nodes (objects) and connectors (associations). Each object has information about itself (attributes). Each object is is defined as a certain type, which determines what attributes it can have and what associations to other objects it can have.

Each metadata object has a unique ID that consists of two parts: the repositoryID and the metadataID. Most metadata objects are stored in the Foundation repository, so the repositoryID is the same for most of the objects that you will be reading.

You can reference an object by its Uniform Resource Identifier (URI), which can have several formats:

omsobj:*ID*
  specifies either the metadataID or the repositoryID.metadataID.

omsobj:*type/ID*
  specifies the metadata type and the ID (for example, omsobj:SASLibrary/repositoryID.metadataID).

omsobj:type?@attribute='value'
  specifies the metadata type and a query of a certain attribute's value (for example, omsobj:SASLibrary?@libref="SASDATA").

All the objects and their attributes and associations are defined in the *SAS Metadata Model: Reference* that is available from the SAS website (support.sas.com).

To look at metadata that is defined in your metadata server, you can launch the Metadata Browser tool from a SAS Display Manager session by going to **Solutions ▶ Accessories ▶ Metadata Browser** or by typing "metabrowse" on a command line.

I find this tool incredibly useful when I'm writing code that will read and write metadata. In the Metadata Browser, you can specify your preferences by going to **Tools ▶ Options ▶ Explorer**. From here you can specify which metadata types to display in the browser (on the **Metadata** tab). If you are very sure of yourself, on the **General** tab, you can change from Browse mode to Edit mode (deselect **Metadata Browse Mode**). You can then directly edit the metadata. I strongly caution against editing the metadata with this method. You can easily break other SAS products and cause the metadata repository to fail.

The following is an example of getting information about a data table from the metadata. I have created metadata using SAS Management Console for a library that has four data tables that are copies of tables from sashelp.

The metadata object that holds the definition of a data table is called a PhysicalTable. A PhysicalTable has an association to a SASLibrary object through the TablePackage association, and it also has associations to zero or more Column objects through the Columns association. So my library as viewed in the Metadata Browser looks like Figure 7.2.

Figure 7.2: Metadata Browser

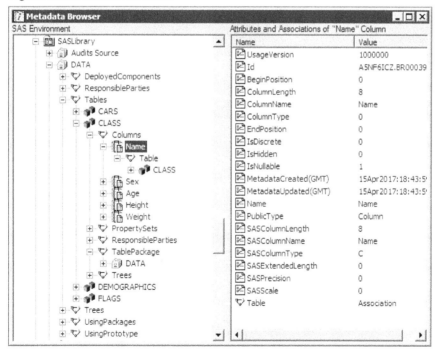

I have expanded the SASLibrary called **DATA**, and it shows all the associations with the ▽ icon. I expanded the **Tables** association, the **CLASS** PhysicalTable object, the **Columns** association, and the **Name** Column object. Clicking the **Name** Column object causes the attributes to be displayed in the right pane. I have also expanded the **Table** association for the **Name** column, and you can see that it refers back to the **CLASS** PhysicalTable. The **TablePackage** association under **CLASS** has been expanded to show that the associated SASLibrary is **DATA**.

If you prefer to visualize this data, Figure 7.3 that shows the objects and their associations.

Figure 7.3: Metadata Objects and Associations.

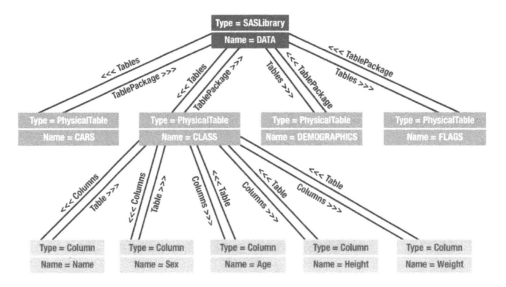

To read the metadata about a library and tables, you can use a DATA step and a set of functions. These are all documented in the Base SAS documentation. Here are the functions that I find the most useful:

metadata_getnobj (uri, n, outuri)
> gets the nth object for this URI. Use this when you are searching for an object with a specific attribute. There might be multiple objects with this attribute, so you can loop through and get all the objects using the n parameter. The object's URI is returned in the outuri variable.

metadata_getnasn (uri, association, n, outuri)
> gets the nth object with the specified association to the given URI. So, using the PhysicalTable object as an example, you can use this function to get each of the Column objects using the Columns association. The object's URI is returned in the outuri variable.

metadata_getattr (uri, attribute, outvalue)
> gets the value of an attribute of an object. The value is returned in the outvalue variable.

Notice that each of these functions returns a value. The last parameter of these functions is a variable, and the function returns the appropriate value in that variable. This means that you must have that variable defined before calling the function, and the value should be set to blank.

There are also metadata functions for writing metadata, but again, you should update metadata only if you are confident that you won't damage anything.

Program 7.9 is an example of reading the metadata for a table to create a table with the correct attributes in the SAS session.

Program 7.9: Reading Metadata to Use When Creating a Data Table

```
options metaserver = localhost    ❶
        metaport = 8561;

%let table = CLASS;
%let library = DATA;

data columns (keep = name type length format label);

   length ptURI    ❷
          libURI
          colURI
          $256;

   length libname $100
          name $40
          type $10
          length $5
          format $20
          label $100;

   ptI = 0;   ❸
   do until (ptURI eq " ");

      ptI + 1;
      ptURI = "";
      rc = metadata_getnobj(
             "omsobj:PhysicalTable?@SASTableName='&table'",
             ptI, ptURI);   ❹

      if (ptURI eq "") then
         leave;   ❺

      libURI = "";    ❻
      rc = metadata_getnasn(ptURI, "TablePackage", 1, libURI);

      if (libURI ne "") then    ❼
      do;
         libname = "";
         rc = metadata_getattr(libURI, "Name", libname);
         if (libname eq "&library") then
            leave;
      end;

   end; /* do until - loop through tables */

   if (ptURI eq "") then
   do;
      putlog "ERR" "OR: Unable to find the table, &table";
      return;
   end;
```

```
    colI = 0;
    do until(colURI eq ""); ❽

        colI + 1;
        colURI = "";
        rc = metadata_getnasn(ptURI, "Columns", colI, colURI);

        if (colURI eq "") then
            leave;

        name = ""; ❾
        rc = metadata_getattr(colURI, "SASColumnName", name);
        label = "";
        rc = metadata_getattr(colURI, "Desc", label);
        type = "";
        rc = metadata_getattr(colURI, "SASColumnType", type);
        length = "";
        rc = metadata_getattr(colURI, "SASColumnLength", length);
        format = "";
        rc = metadata_getattr(colURI, "SASFormat", format);

        output;

    end; /* do until - loop through columns */
run;
```

❶　Connect to the metadata server and specify the metadata name of the table and library that you are looking for.

❷　Set up the URI variables. These must be defined before using them in the metadata functions. Set up the variables that are used to get the attribute values. These also must be defined before using them in the metadata functions.

❸　Set up a counter to iterate through the PhysicalTables.

❹　Get each PhysicalTable with &table as the name.

❺　If the URI returned by the metadata_getnobj function is blank, then all the table metadata has been read, so stop processing.

❻　Get the library for this table using the TablePackage association.

❼　If there is a library, then get its name and see whether it matches &library. If it does match, then we have found the correct table and library, so stop iterating through tables.

❽　Iterate through the columns.

❾　Get the attributes from the column object.

After you have retrieved the information from metadata, you can use the code in Program 7.10 to create the tables. SAS Log 7.5 shows the generated code.

Program 7.10: Creating a Table Using Metadata

```
data _null_;
   set columns end = eof;
   if (_n_ eq 1) then
    call execute("data &table;");
   if (type eq "C") then
      length = "$" !! length;
   call execute("  attrib " !! strip(name));
   call execute("    length = " !! strip(length));
   if (format ne "") then
      call execute("    format = " !! strip(format));
   if (label ne "") then
      call execute("    label = '" !! strip(label) !! "'");
   call execute("  ;");
   if (eof) then
   do;
      call execute("  call missing(of _all_);");
      call execute("  stop;");
      call execute("run;");
   end;
run;
```

SAS Log 7.5: Creating a Table Using Metadata

```
NOTE: CALL EXECUTE generated line.
1    + data CLASS;
2    +     attrib Name
3    +             length = $8
4    +             format = $char8.
5    +             label = 'Student Name'
6    +     ;
7    +     attrib Sex
8    +             length = $1
9    +             format = $char1.
10   +             label = 'Sex'
11   +     ;
12   +     attrib Age
13   +             length = 8
14   +             label = 'Age'
15   +     ;
16   +     attrib Height
17   +             length = 8
18   +             format = 3.
19   +             label = 'Height (inches)'
20   +     ;
21   +     attrib Weight
22   +             length = 8
23   +             format = 4.1
24   +             label = 'Weight (pounds)'
25   +     ;
26   +     call missing(of _all_);
27   +     stop;
28   + run;
```

```
NOTE: The data set WORK.CLASS has 0 observations and 5 variables.
NOTE: DATA statement used (Total process time):
      real time          0.01 seconds
      cpu time           0.01 seconds
```

Using the XML LIBNAME Engine

I recently developed an application that has several XML files that the user can modify to configure the system. In my SAS programs, I read the XML files using the XML LIBNAME Engine, which enables me to access the configuration data as SAS tables. I can then use these tables to set macro variables and options and do other configuration tasks.

For the following example of the XML LIBNAME Engine, I have a simplified configuration file, shown in XML File 7.1, that has some macros variables that I want to create, some LIBNAME statements to create, some options to set, and a list of the valid users of the system and their roles. Note that this XML format is my own creation. The XML engine works with most XML files.

XML File 7.1: Configuration XML file

```
<config>

  <macro name="admin" value="sasadmin"/>
  <macro name="password" value="{SAS002}83FBCB5C4121449726F0A94F"/>
  <macro name="dev" value="devSystem"/>
  <macro name="prod" value="prodSystem"/>

  <library libref="wh" path="c:\project\warehouse"/>
  <library libref="formats" path="c:\project\formats"/>

  <option name="sasautos" value="'c:\project\sasautos'"
                          insert="yes"/>
  <option name="fmtsearch" value="formats" insert="yes"/>
  <option name="SOURCE2"/>
  <option name="minoperator"/>
  <option name="mindelimiter" value="','"/>

  <user userid="user1" name="First User">
    <role name="ETL"/>
    <role name="REPORTING"/>
  </user>
  <user userid="user2" name="Second User">
    <role name="REPORTING"/>
  </user>

</config>
```

If you are not familiar with XML, here is a very brief overview to get you started.

XML has tags and attributes and values. The tags are enclosed in angle brackets (< >), like this:

```
<tag>
```

So in XML File 7.1, <config>, <macro>, <library>, <option>, <user>, and <role> are all tags. You must close a tag by putting a / before the closing bracket:

```
<tag/>
```

Or you can use a closing tag like this:

```
<tag>
</tag>
```

You can also have subtags:

```
<tag>
  <subtag>
    <subtag2/>
  </subtag>
</tag>
```

Within the brackets, you can specify attributes, which are in the format of attribute="value":

```
<tag attribute="value">
```

You can also specify values for the tag:

```
<tag>value</tag>
```

You cannot have a subtag and a value. If you need to do this, then use an attribute instead of a value.

The SAS XML LIBNAME engine displays XML as SAS tables. It makes a SAS data table for each tag that it finds, with columns created from the attributes and the values. So when the XML engine is used with XML File 7.1, the library creates a table called macros with an admin variable and a name variable, as well as various other tables and columns. When you have subtags, you need to be able to connect them to the parent tags, so the LIBNAME engine creates key variables (called "ordinals") that enable you to join subtag tables with parent tag tables. In XML File 7.1, the user table and the role table each have an ordinal variable that enables the role records to match the appropriate user records.

To use the XML engine to read this file, you need an XML map file that tells the engine which XML elements to map to which tables. The XML map file is another piece of XML that has <TABLE> and <COLUMN> tags that describe the mappings. Luckily, you don't have to create the map file from scratch; the SAS XML Mapper tool can create it for you. This tool can be downloaded from support.sas.com: **Support ▶ Downloads & Hot Fixes ▶ SAS System Software ▶ Base SAS Software**.

After you have the SAS XML Mapper tool installed, open it and use **File ▶ Open XML** to open the XML file. If you have an XML schema (a definition file that describes the XML structure), you can open it instead. Then use **Tools ▶ AutoMap using XML** to create the map. Figure 7.4 shows the SAS XML Mapper with my XML file loaded into it.

Figure 7.4: SAS XML Mapper

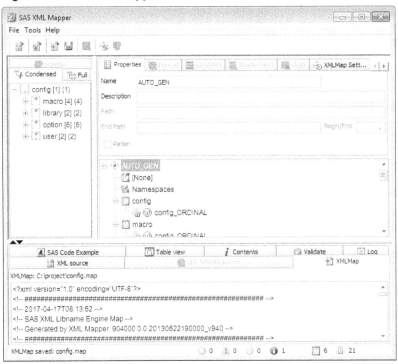

From here you can modify the properties of the map in the tabbed section in the upper right, or you can wait and modify the map file after it is created. Use **File ▶ Save XML Map** to create the map file.

When you open the map file, you see the <TABLE> and <COLUMN> tags that define the SAS tables and columns. In addition to the SAS columns that are created from the XML attributes and values, there are columns called xxx_ORDINAL, which are key fields that enable you to join interconnected tables. In my example, the role table contains the user_ORDINAL that points to the user table, so that the roles can be combined with the users. You can modify the map XML if you want. I usually get rid of ORDINAL and TABLE tags that I don't need and make sure all my column attributes are correct.

The map file that I ended up with (after modifying the output from the SAS XML Mapper) looks like XML File 7.2.

XML File 7.2: XML Map File

```
<?xml version="1.0" encoding="UTF-8"?>
<!-- ############################################################## -->
<!-- SAS XML Libname Engine Map -->
<!-- Generated by XML Mapper, 904000.0.0.20130522190000_v940 -->
<!-- ############################################################## -->
<!-- ### Validation report                  ### -->
<!-- ############################################################## -->
<!-- XMLMap validation completed successfully. -->
<!-- ############################################################## -->
```

```
<SXLEMAP name="AUTO_GEN" version="2.1">
    <NAMESPACES count="0"/>

    <!-- ##################################################### -->
    <TABLE description="macro" name="macro">  ❶
        <TABLE-PATH syntax="XPath">/config/macro</TABLE-PATH>  ❷
        <COLUMN name="name">
            <PATH syntax="XPath">/config/macro/@name</PATH>  ❸
            <TYPE>character</TYPE>
            <DATATYPE>string</DATATYPE>  ❹
            <LENGTH>8</LENGTH>
        </COLUMN>
        <COLUMN name="value">
            <PATH syntax="XPath">/config/macro/@value</PATH>
            <TYPE>character</TYPE>
            <DATATYPE>string</DATATYPE>
            <LENGTH>32</LENGTH>
        </COLUMN>
    </TABLE>

    <!-- ##################################################### -->
    <TABLE description="library" name="library">
        <TABLE-PATH syntax="XPath">/config/library</TABLE-PATH>
        <COLUMN name="libref">
            <PATH syntax="XPath">/config/library/@libref</PATH>
            <TYPE>character</TYPE>
            <DATATYPE>string</DATATYPE>
            <LENGTH>7</LENGTH>
        </COLUMN>
        <COLUMN name="path">
            <PATH syntax="XPath">/config/library/@path</PATH>
            <TYPE>character</TYPE>
            <DATATYPE>string</DATATYPE>
            <LENGTH>20</LENGTH>
        </COLUMN>
    </TABLE>

    <!-- ##################################################### -->
    <TABLE description="option" name="option">
        <TABLE-PATH syntax="XPath">/config/option</TABLE-PATH>
        <COLUMN name="name">
            <PATH syntax="XPath">/config/option/@name</PATH>
            <TYPE>character</TYPE>
            <DATATYPE>string</DATATYPE>
            <LENGTH>12</LENGTH>
        </COLUMN>
        <COLUMN name="value">
            <PATH syntax="XPath">/config/option/@value</PATH>
            <TYPE>character</TYPE>
            <DATATYPE>string</DATATYPE>
            <LENGTH>25</LENGTH>
        </COLUMN>
        <COLUMN name="insert">
            <PATH syntax="XPath">/config/option/@insert</PATH>
            <TYPE>character</TYPE>
```

```
                <DATATYPE>string</DATATYPE>
                <LENGTH>3</LENGTH>
            </COLUMN>
        </TABLE>

        <!-- ######################################################## -->
        <TABLE description="user" name="user">
            <TABLE-PATH syntax="XPath">/config/user</TABLE-PATH>
            <COLUMN class="ORDINAL" name="user_ORDINAL">  ❺
                <INCREMENT-PATH beginend="BEGIN"
                 syntax="XPath">/config/user</INCREMENT-PATH>
                <TYPE>numeric</TYPE>
                <DATATYPE>integer</DATATYPE>
            </COLUMN>
            <COLUMN name="userid">
                <PATH syntax="XPath">/config/user/@userid</PATH>
                <TYPE>character</TYPE>
                <DATATYPE>string</DATATYPE>
                <LENGTH>5</LENGTH>
            </COLUMN>
            <COLUMN name="name">
                <PATH syntax="XPath">/config/user/@name</PATH>
                <TYPE>character</TYPE>
                <DATATYPE>string</DATATYPE>
                <LENGTH>11</LENGTH>
            </COLUMN>
        </TABLE>

        <!-- ######################################################## -->
        <TABLE description="role" name="role">
            <TABLE-PATH syntax="XPath">/config/user/role</TABLE-PATH>
            <COLUMN class="ORDINAL" name="user_ORDINAL">   ❺
                <INCREMENT-PATH beginend="BEGIN"
                 syntax="XPath">/config/user</INCREMENT-PATH>
                <TYPE>numeric</TYPE>
                <DATATYPE>integer</DATATYPE>
            </COLUMN>
            <COLUMN name="name">
                <PATH syntax="XPath">/config/user/role/@name</PATH>
                <TYPE>character</TYPE>
                <DATATYPE>string</DATATYPE>
                <LENGTH>9</LENGTH>
            </COLUMN>
        </TABLE>

</SXLEMAP>
```

❶ The <TABLE> tag defines the table. The name attribute is the name of the SAS table, and the description attribute is the data table's label.

❷ The <TABLEPATH> tag describes which tag to look for to fill the table. In this case, that path is /config/macro, so the data comes from any <macro> subtag of the <config> tag. Note that the <config> tag is the parent tag of everything in the XML.

❸ Each <COLUMN> tag has a <PATH> tag that defines where to read the value. The <PATH> value of /config/macro/@name indicates that the values come from the name attribute of the

<macro> subtag of the <config> tag. The @ indicates that this is an attribute. If <macro> had a value, then the path would simply be /config/macro.

❹ Also included under the <COLUMN> tag are tags for configuring the SAS variable that is created.

❺ Because the <role> tags are subtags of the <user> tags, you need key values to connect the roles with the right users. These are the ORDINAL variables. The class attribute on the <COLUMN> tag tells the engine that this is an ordinal, and then the <INCREMENT-PATH> tag tells it which tag to count. The engine increments the ordinal value each time it finds a new tag. In this case, the path is /config/user, so the ordinal is incremented each time a new <user> subtag of the <config> tag is found.

When the XML file and the XML map file are ready, the code in Program 7.11 can be used to read the XML and create the macros, LIBNAME statements, and options that are specified. It also creates a table that contains the valid users and their roles, which is shown in Figure 7.5.

Program 7.11: Reading the XML Configuration File

```
libname syscfg xmlv2 "c:\project\config.xml"
          xmlmap="c:\project\config.map";  ❶

data _null_;
   set syscfg.macro;  ❷
   call symputx(cats("_", name), strip(value), "G");
run;

data _null_;
   set syscfg.library;  ❸
   rc = libname(libref, path);
run;

data _null_;
   set syscfg.option;  ❹
   if (value eq "") then
      call execute("option " !! strip(name) !! ";");
   else if (insert eq "") then
      call execute("option " !! strip(name) !! " = " !!
                                 strip(value) !! ";");
   else
      call execute("option insert=(" !! strip(name) !! " = " !!
                                     strip(value) !! ");");
run;

proc sql;  ❺
   create table users as
      select u.userid, u.name, r.name as role
         from syscfg.user as u
              left join
              syscfg.role as r
                on u.user_ORDINAL eq r.user_ORDINAL;
quit;

libname syscfg;
```

❶ Issue LIBNAME using the xmlv2 engine, point to the XML file, and specify the XML map file.

❷ Read the config macro table and create global macro variables.

❸ Read the config library table and create LIBNAME statements.

❹ Read the config option table and create OPTION statements.

❺ Join the config user table with the config role table using the user_ORDINAL field.

Figure 7.5: users Table

Obs	userid	name	role
1	user1	First User	ETL
2	user1	First User	REPORTING
3	user2	Second User	REPORTING

Appendix A: Utility Macros

Introduction

I have a set of utility macros that I usually have in my SASAUTOS library. These macros simplify tasks that I do regularly. This appendix provides documentation about each macro.

I use a number of standards when writing my utilities to keep them from conflicting with other programs:

- The macro names always start with an underscore.

- I use l_ at the beginning of all my local macro variable names, and I define them with a %LOCAL statement so that I don't risk deleting any macro variables in the calling program.

- I create and use some global macro variables, and I begin the names of these macro variables with two underscores so that they won't interfere with the rest of the environment.

- If I make temporary tables in the macro, their names begin with tmp_, and they are deleted at the end of the macro.

- Many of my utility and other macros use a debug flag variable that I call &__DEBUG. If &__DEBUG is blank or 0, then no debugging takes place. Values that are greater than zero mean that some debugging takes place. You can use a %LET statement to set this macro variable to anything greater than zero to turn on debugging before calling the utility macro.

- If there is only one parameter, I use a positional parameter (with no equal sign). If there is more than one parameter, then I use named parameters (with equal signs).

- I don't call any other custom macros from within these macros, because nothing is more annoying than trying to use a macro and finding out that you also need five other macros to run it.

Some of these utilities are macro functions rather than standard macros. This means that they return a value to the calling program, so they must be called from a place where the return of a value makes sense. Generally this means that they are used in a %LET statement, a %IF statement, a DATA step assignment, or a DATA step IF statement.

Deleting Tables

The %_delete_tables utility will delete multiple tables from more than one library in a single macro call. It is useful for keeping programs from leaving temporary tables behind. It runs PROC DATASETS to delete the specified tables. The NOWARN option prevents warnings in the SAS log, even if a table doesn't exist.

If the &__DEBUG global macro variable is greater than 0, then the tables are not deleted. This is helpful when you are trying to figure out why your program isn't doing what it should. You just set &__DEBUG to 1 before calling running the program, and the %_delete_tables macro calls do not delete your tables.

Syntax

The syntax is as follows:

```
%_delete_tables(tableList)
```

Parameters

tableList
    Specifies a list of data tables. The table names should be delimited with either a space or a comma. The table names can be one-level names (for WORK tables), two-level names, or a mixture of both.

Example

The example in Program A.1 and SAS Log A.1 shows an attempt to delete two WORK tables and a SASUSER table. The WORK.files table does not exist, so it is ignored. The log reports which tables were deleted.

Program A.1: Running %_delete_tables

```
%_delete_tables(test, files, sasuser.test);
```

SAS Log A.1: Running %_delete_tables

```
1    %_delete_tables(test, files, sasuser.test);

NOTE: Deleting SASUSER.TEST (memtype=DATA).
NOTE: PROCEDURE DATASETS used (Total process time):
      real time           0.01 seconds
      cpu time            0.01 seconds

NOTE: Deleting WORK.TEST (memtype=DATA).
NOTE: PROCEDURE DATASETS used (Total process time):
      real time           0.68 seconds
      cpu time            0.67 seconds
```

Getting the Number of Records

The %_get_num_records macro function returns the number of records (active records, not records marked as deleted) in a table. If the table doesn't exist, then it returns -1.

This function is very useful if you want to skip some code when a table doesn't exist or doesn't have any data: You can use this macro in a macro %IF statement like this:

```
%if (%_get_num_records(work.table) gt 0) %then...
```

Or you can use it in a %LET statement:

```
%let nobs = %_get_num_records(work.table);
```

It can also be used from a DATA step:

```
data _null_;
   nobs = %_get_num_records(work.table);
   putlog nobs=;
   if (%_get_num_records(work.otherTable) eq -1) then
      putlog "otherTable does not exist";
run;
```

The macro also accepts a table with a WHERE clause and returns the number of records in the subset.

Syntax

The syntax is as follows:

```
%_get_num_records(table)
```

Parameters

table

Specifies the name of the table from which to get the number of records. This can be a one-level (WORK) or a two-level table name, and it can include a WHERE clause in the following form: tableName(where=(where_clause)).

Return

The function returns the following values:

−1   indicates that the table does not exist.

0   indicates that the table exists but has no records.

>0   indicates the number of records in the table.

Example

In the Program A.2 and SAS Log A.2 example, the first %LET statement shows getting the records for a table, and the second %LET statement shows getting the records from a table with a WHERE clause applied to it.

Program A.2: Running %_get_num_records

```
%let nobs = %_get_num_records(sashelp.class);
%put &=nobs;

%let nobs = %_get_num_records(
              %str(sashelp.class(where=(sex = "F"))));
%put &=nobs;
```

SAS Log A.2: Running %_get_num_records

```
1    %let nobs = %_get_num_records(sashelp.class);
2    %put &=nobs;
NOBS=19
3
4    %let nobs = %_get_num_records(%str(sashelp.class(where=(sex =
"F"))));
5    %put &=nobs;
NOBS=9
```

Getting a Library's Engine Name

The %_get_engine macro function returns the name of the engine of a library. This function is useful if you need to do different processing based on the type of engine (for example, you might need to use pass-through if this is a DBMS engine).

Syntax

The syntax is as follows:

```
%_get_engine(libref)
```

Parameters

libref
    Specifies the libref that points to a library defined in your SAS session.

Return

The function returns the name of the engine of the library (for example, V9 = the Base SAS engine for a SAS 9 library).

The function returns a blank if the libref hasn't been defined.

Example

The example in Program A.3 and SAS Log A.3 shows how to get the engine name from a Base SAS library and from an XML library. It also attempts to get an engine name from a library that hasn't been defined.

Program A.3: Running %_get_engine

```
%let engine = %_get_engine(sashelp); /* base sas library */
%put &=engine;

%let engine = %_get_engine(syscfg);  /* xml engine library */
%put &=engine;

%let engine = %_get_engine(xyz);     /* xyz doesn't exist */
%put &=engine;
```

SAS Log A.3: Running %_get_engine

```
8    %let engine = %_get_engine(sashelp); /* base sas library */
2    %put &=engine;
ENGINE=V9
3
4    %let engine = %_get_engine(syscfg);  /* xml engine library */
5    %put &=engine;
ENGINE=XMLV2
6
7    %let engine = %_get_engine(xyz);     /* xyz doesn't exist */
8    %put &=engine;
ENGINE=
```

Getting a Variable Keep List

The %_get_keeplist macro function gets a list of variable names from the specified table and returns the list in the format needed—either as a keep list (with spaces between the variables) or for a PROC SQL select clause (with an alias before each variable name and commas between variables). If you specify a PROC SQL alias, then the select clause style is returned. Otherwise, the standard keep list is returned.

Syntax

The syntax is as follows:

```
%_get_keeplist(table=, sqlAlias=)
```

Parameters

table

Specifies the name of the table to get the keep list from. This can be a one-level (WORK) or two-level name. This parameter is required.

sqlAlias

Specifies a SQL alias if you want a list of variable names for use in a PROC SQL SELECT statement. The alias is specified on the PROC SQL FROM statement: from table as sqlAlias.

Return

The function returns a list of variables either in the format for a KEEP statement, where the names are delimited with spaces, or in the format for a PROC SQL SELECT statement, where the names are delimited by commas and each name is prefixed with sqlAlias.

If the table doesn't exist or has no variables, then a blank is returned.

Example

Program A.4 and SAS Log A.4 show how to get a list of variables to use in a KEEP statement.

Program A.4: Running %_get_keeplist for a KEEP Statement

```
%let keep = %_get_keeplist(table = sashelp.class);
data new (keep = &keep);
   set sashelp.classfit;
run;
```

SAS Log A.4: Running %_get_keeplist for a KEEP Statement

```
1    data new
2 !  (keep = &keep);
SYMBOLGEN:  Macro variable KEEP resolves to Name Sex Age Height
Weight
3        set sashelp.classfit;
4    run;
```

Program A.5 and SAS Log A.5 show how to get a list of variables for a SELECT statement.

Program A.5: Running %_get_keeplist for a SELECT Statement

```
proc sql;
   select %_get_keeplist(table = sashelp.class, sqlAlias = t),
          "test" as newVar
       from sashelp.class as t;
quit;
```

SAS Log A.5: Running %_get_keeplist for a SELECT Statement

```
1    proc sql;
2        select %_get_keeplist(table = sashelp.class, sqlAlias = t),
MPRINT(_GET_KEEPLIST):   t.Name, t.Sex, t.Age, t.Height, t.Weight
3               "test" as newVar
4           from sashelp.class as t;
5    quit;
```

Making ATTRIB Statements

The %_make_attribs macro can be called from inside a DATA step to create ATTRIB statements based on the variables in a data table. If the table doesn't exist or has no variables, then no statements are created, and a note is displayed in the SAS log.

Syntax

The syntax is as follows:

```
%_make_attribs(table);
```

Parameters

table
Specifies the name of the table from which to get the variables and attributes. This parameter is required.

Example

Program A.6 and SAS Log A.6 show how to use the macro to create the ATTRIB statements based on the variables in sashelp.class.

Program A.6: Running %_make_attribs

```
options mprint;
data new;
   %_make_attribs(sashelp.class);
   call missing(of _all_);
run;
```

SAS Log A.6: Running %_make_attribs

```
1    options mprint;
2    data new;
3        %_make_attribs(sashelp.class);
MPRINT(_MAKE_ATTRIBS):   attrib Name length = $8 ;
MPRINT(_MAKE_ATTRIBS):   attrib Sex length = $1 ;
MPRINT(_MAKE_ATTRIBS):   attrib Age length = 8 ;
MPRINT(_MAKE_ATTRIBS):   attrib Height length = 8 ;
MPRINT(_MAKE_ATTRIBS):   attrib Weight length = 8 ;
4        call missing(of _all_);
5    run;
```

Making a Basic Format

The %_make_format macro creates a format from a data table.

Syntax

The syntax is as follows:

```
%_make_format(inputTable=,
              fmtLib=,
              fmtName=,
```

```
              start=,
              label=,
              otherLabel=,
              type=);
```

Parameters

inputTable
> Specifies the name of the table to use as input. This parameter is required.

fmtLib
> Specifies the library in which to store the format. If blank, then it defaults to WORK.

fmtName
> Specifies the name of the format to be created. This parameter is required.

start
> Specifies the variable that holds the start value. This parameter is required.

label
> Specifies the variable that holds the label value. This parameter is required.

otherLabel
> Specifies the value to use for "other" values. If blank, no "other" value is created.

type
> Specifies the type of format: C (character format), N (numeric informat), J (character informat), or I (numeric informat). If blank, then it defaults to C.

Example

Program A.7 creates a format from the sashelp.class table that will convert a name value to a sex value. It sets the "other" value to "X", so any names that are not in the class table are set to "X". The DATA step tests the format. "William" and "Jane" are both in the format, and "Martha" is not, so "William" has a sex of "M", "Jane" is "F", and "Martha" is "X". The output is shown in SAS Log A.7.

Program A.7: Running %_make_format

```
%_make_format(inputTable = sashelp.class,
              fmtName = $sex,
              start = name,
              label = sex,
              otherLabel = 'X');

data _null_;
   n = "William";
   s = put(n, $sex.);
   putlog n= s=;
   n = "Jane";
   s = put(n, $sex.);
   putlog n= s=;
   n = "Martha";
   s = put(n, $sex.);
   putlog n= s=;
run;
```

SAS Log A.7: Running %_MAKE_FORMAT%_make_format

```
n=William s=M
n=Jane s=F
n=Martha s=X
```

Making a Directory Path

The %_make_directory macro creates the directories in the specified path. It can be used on either Windows or Unix. This macro will check each level of the supplied path and will create the levels that do not already exist. If the entire path already exists, nothing is created.

Syntax

The syntax is as follows:

```
%_make_directory(path);
```

Parameters

path

> specifies the full path to be created. Windows path are expected in the drive:\level1\level2 format, and Unix paths are expected in the /level1/level2 format. This parameter is required.

Example

Program A.8 is run on Windows, and none of the path levels exist before running the macro. Figure A.1 shows the directories that are created.

Program A.8: Running %_make_directory on Windows

```
%_make_directory(c:\one\two\three);
```

Figure A.1: Running %_make_directory on Windows

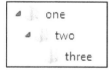

Program A.9 is run on Unix, and the first 2 levels of the path already exist before running the macro. Figure A.2 shows the directories on Unix after running the macro.

Program A.9: Running %_make_directory on Unix

```
%_make_directory(/home/user/one/two/three);
```

Figure A.2: Running %_make_directory on Unix

```
|-home
|---user
|-----one
|-------two
|---------three
```

Creating Macro Variables from SYSPARM

If you use the SYSPARM option to send values into a SAS program, then you can use the %_read_sysparm macro to read the contents of SYSPARM and create global macro variables from the contents. The macro assumes that the contents are space- or comma-delimited, and the values are either simple values or in name-value pairs (name=value). If just a value is found, then the macro variable is called parm#. An additional macro variable, __macroCount, is created to tell you how many parm# variables were created. Each time a macro variable is created, a note is written to the SAS log with the macro variable name and value.

Syntax

The syntax is as follows:

```
%_read_sysparm(notes)
```

Parameters

notes
    Set this parameter to nonotes to turn off the macro variable name and value notes.

Example

Program A.10 sets the SYSPARM option (this option is normally set when you start SAS, but can also be set with an OPTIONS statement), and then calls the utility macro to read the values. It then puts out the values of the parameters and the &__macroCount variable. The SYSPARM values are separated by commas and spaces. Some of the values are name-value pairs, and some are just values. The output from the program is shown in SAS Log A.8.

Program A.10: Running %_ read_sysparm

```
options sysparm = "a=1 b=2, 3, 4, c=5";
%_read_sysparm ();
%put &=__macroCount;
```

SAS Log A.8: Running %_read_sysparm

```
1    options sysparm = "a=1 b=2, 3, 4, c=5";
2    %_read_sysparm ();
Macro variable created: &a = 1
Macro variable created: &b = 2
Macro variable created: &parm1 = 3
Macro variable created: &parm2 = 4
Macro variable created: &c = 5
3    %put &=__macroCount;
__MACROCOUNT=2
```

Setting Log Options

The %_turn_log_options_off, %_turn_log_options_on, and %_reset_log_options macros work together to turn on and off the system options that control what is displayed in the SAS log.

If you want to run some code that doesn't write anything to the SAS log, you can use the %_turn_log_options_off macro to turn the following options off: SOURCE, SOURCE2, NOTES,

MLOGIC, MLOGICNEST, MPRINT, MPRINTNEST, and SYMBOLGEN. The macro saves the original values of these options in global macro variables so that the options can be reset to their original values when you want to turn them on again. The %_reset_log_options macro uses those global macro variables to reset the options for you. So you can call %_turn_log_options_off at the beginning of a program, and then call %_reset_log_options at the end of the program to get almost nothing in the log from running the program.

For debugging, the %_turn_log_options_on macro sets all the log options on so that you get as much log information as SAS can provide. Because this macro also saves the original option settings in macro variables, you can use the %_reset_log_options macro to restore your standard settings.

> **Tip:** You can save all options and their values in a SAS table by using PROC OPTSAVE, and you can reload them using PROC OPTLOAD.
>
> I don't use these procedures in my utility macros because I try to avoid as much "noise" in the SAS log as possible. But these procedures are great in other cases to save and reset options.

Syntax

The syntax for the three macros is as follows:

```
%_turn_log_options_off
%_tun_log options_on
%_reset_log_options
```

Parameters

There are no parameters for these macros.

Example

Program A.11 shows a test macro that has some macro statements, followed by a DATA step. SAS Log A.9 shows the output when the macro is run using the default option settings in a SAS session.

Program A.11: Standard Output from a Macro

```
%macro test;
    %let x = 1;
    %if (&x eq 1) %then
    %do;
        data _null_;
            putlog "In data step";
        run;
        %put mprint:    %sysfunc(getoption(mprint));
        %put mlogic:    %sysfunc(getoption(mlogic));
        %put symbolgen: %sysfunc(getoption(symbolgen));
        %put source:    %sysfunc(getoption(source));
        %put notes:     %sysfunc(getoption(notes));
    %end;
%mend test;
%test;
```

SAS Log A.9: Standard Output from a Macro

```
15  %test
In data step
NOTE: DATA statement used (Total process time):
      real time               0.00 seconds
      cpu time                0.00 seconds


mprint:     NOMPRINT
mlogic:     NOMLOGIC
symbolgen: NOSYMBOLGEN
source:     SOURCE
notes:      NOTES
```

Program A.12 runs %_turn_log_options_on and then the same %TEST macro, and SAS Log A.10 shows the output that is created.

Program A.12: Running %_turn_log_options_on

```
%_turn_log_options_on;
%test;
```

SAS Log A.10: Running %_turn_log_options_on

```
17  %test;
MLOGIC(TEST):  Beginning execution.
MLOGIC(TEST):  %LET (variable name is X)
SYMBOLGEN:  Macro variable X resolves to 1
MLOGIC(TEST):  %IF condition (&x eq 1) is TRUE
MPRINT(TEST):   data _null_;
MPRINT(TEST):   putlog "In data step";
MPRINT(TEST):   run;

In data step
NOTE: DATA statement used (Total process time):
      real time               0.00 seconds
      cpu time                0.00 seconds


MLOGIC(TEST):  %PUT mprint:    %sysfunc(getoption(mprint))
mprint:     MPRINT
MLOGIC(TEST):  %PUT mlogic:    %sysfunc(getoption(mlogic))
mlogic:     MLOGIC
MLOGIC(TEST):  %PUT symbolgen: %sysfunc(getoption(symbolgen))
symbolgen: SYMBOLGEN
MLOGIC(TEST):  %PUT source:    %sysfunc(getoption(source))
source:     SOURCE
MLOGIC(TEST):  %PUT notes:     %sysfunc(getoption(notes))
notes:      NOTES
MLOGIC(TEST):  Ending execution.
```

Program A.13 runs %_turn_log_options_off and then the same %TEST macro, and SAS Log A.11 shows the output that is created.

Program A.13: Running %_turn_log_options_off

```
%_turn_log_options_off;
%test;
```

SAS Log A.11: Running %_turn_log_options_off

```
In data step
mprint:     NOMPRINT
mlogic:     NOMLOGIC
symbolgen:  NOSYMBOLGEN
source:     NOSOURCE
notes:      NONOTES
```

Finally, Program A.14 runs %_RESET_LOG_OPTIONS and then the same %test macro, and SAS Log A.12 shows that the output is back to the original output.

Program A.14: Running %_reset_log_options

```
%_reset_log_options;
%test;
```

SAS Log A.12: Running %_reset_log_options

```
21   %test;

In data step
NOTE: DATA statement used (Total process time):
      real time           0.00 seconds
      cpu time            0.00 seconds

mprint:     NOMPRINT
mlogic:     NOMLOGIC
symbolgen:  NOSYMBOLGEN
source:     SOURCE
notes:      NOTES
```

Displaying Macro Notes for Debugging

The %_debug_macro_note macro writes a note in the log with the name of the macro and a list of the parameters and their values. I call this macro at the beginning of all the non-utility macros in my systems. This is just for debugging, so it only writes notes if the &__DEBUG variable is set to a number greater than 0.

Do not call this macro from a macro function; the function will fail.

Syntax

The syntax is as follows:

```
%_debug_macro_note
```

Parameters

There are no parameters for this macro.

Example

Program A.15 and SAS Log A.13 show that the utility macro does nothing when &__DEBUG is set to 0.

Program A.15: Running %_debug_macro_note with &__DEBUG Set to 0

```
%macro test(a=, b=);
   %_debug_macro_note;
   %put In test macro;
%mend test;

%let __debug = 0;
%test (a = 1, b = 2);
```

SAS Log A.13: Running %_debug_macro_note with &__DEBUG Set to 0

```
7    %test (a = 1, b = 2);
In test macro
```

Program A.16 and SAS Log A.14 show the note that is displayed in the SAS log.

Program A.16: Running %_debug_macro_note with &__DEBUG Set to 1

```
%let __debug = 1;
%test (a = 1, b = 2);
```

SAS Log A.14: Running %_debug_macro_note with &__DEBUG Set to 1

```
9    %test (a = 1, b = 2);
DEBUG: =====
DEBUG: ===== Macro: TEST
DEBUG: ===== Parameters:
DEBUG: =====     &A = 1
DEBUG: =====     &B = 2
DEBUG: =====
In test macro
```

Tip: The code for this macro is a good example of figuring out where you are in the macro stack.

Refreshing Autocall Macros

The %_refresh_macros macro can be run at any time to include all the SASAUTOS macros into the SAS session and submit them for you. It includes both external files and sas catalog source members.

This is a development macro. It is not meant to be used in a production process. It is meant to be used in a SAS session where you have a SASAUTOS library. If any of the macros are changed outside of the SAS session, then you have to include the updated macro into the SAS session and submit it in order to get the latest version. I run into this situation quite a bit when I'm working on

Unix, because I prefer to use an editor other than the SAS Program Editor. Note that the macro does not include and submit macros that are part of the SAS system.

The macro uses the &__DEBUG macro variable. If it is blank or 0, then nothing is added to the SAS log (other than a note that tells you that the macros have been refreshed). If you set &__DEBUG to 1 or more, you get lots of information in the SAS log.

Syntax

The syntax is as follows:

```
%_refresh_macros;
```

Parameters

There are no parameters for this macro.

Example

To use the macro, just call it from a SAS session as in Program A.17. SAS Log A.15 shows that only a note is displayed in the SAS log.

Program A.17: Running %_refresh_macros

```
%_refresh_macros;
```

SAS Log A.15: Running %_refresh_macros

```
1   %_refresh_macros;
===> SASAUTO macros have been refreshed
```

Appendix B: Display Manager

Introduction

There are many ways to write and run SAS code—SAS Enterprise Guide, SAS Studio, random text editors, and batch processing. I'm still old-fashioned enough to use the SAS Display Manager, also called DMS, the SAS Windowing Environment, or just Base SAS. It is the windowing environment that you get when you run the sas executable: it has multiple windows, including an editor window, a log window, an explorer window, and an output window. If you are running on a Windows operating system, then all of the SAS windows are contained within a single window. If you are running on a Unix operating system, then the SAS windows are displayed as separate windows. If you can avoid using DMS on a z/OS system, you should; it is not very easy to use.

If you use the Display Manager, then I have some tips that can make you more productive. If you have access to SAS Enterprise Guide or SAS Studio and prefer them, then skip this section.

Using the Enhanced Editor

If you are running SAS on Windows, you can use the Enhanced Editor. It is a much more flexible editor than the original Program Editor, and you can open multiple Enhanced Editor windows in DMS, while there is only one Program Editor window.

Indenting Your Code

I'm a strong believer in using white space to improve code readability, so indenting is important to me. You can use the following techniques:

Indent
> If you have a section of code that needs to be indented, then select the section and use the Tab key to indent all the lines in that section.

Unindent
> To unindent a section, select the code and use Shift+Tab to unindent all the lines in that section.

The size of the indent is specified in the **Tools ▶ Options ▶ Enhanced Editor ▶ General** window.

Commenting Out a Section of Code

Sometimes there is a section of code that you want to comment out, either temporarily or permanently. You can use the following techniques:

Comment
> To add a /* at the beginning of each line and a */ at the end of each line, select a section of code and use Ctrl+/

Uncomment
> If you have a section that already has /* */ around each line (starting in the first column), you can select the block and use Ctrl+Shift+/

Converting a String to Uppercase or Lowercase

If you need to convert a string in your code to uppercase or lowercase, use these techniques:

Uppercase
> Select the string and use Ctrl+U to convert it to uppercase.

Lowercase
> Select the string and use Ctrl+L to convert it to lowercase.

Removing Excessive White Space

Sometimes you can end up with extra blanks at the end of a line of code. These extra blanks make the saved program bigger and can also make the horizontal scroll bar on the editor indicate that there is more code to the right when there are actually only blanks. To get rid of all the spaces at the end of the line, select the lines that you want to remove the white space from, and use Ctrl+Shift+W.

Opening a New Enhanced Editor Window

If you are currently in an Enhanced Editor window and you want to open a new empty editor window, you can click the New icon at the top of the window. The New icon creates a new Enhanced Editor window only if you click it when you are in an Enhanced Editor window.

I sometimes find that I've closed all my Enhanced Editor windows, so there is no icon to click to bring up a new one. The only way to get a new window is to go to **View ▶ Enhanced Editor**.

To avoid the extra click, I have customized the toolbars for some of the other windows (the log and explorer windows are the ones I mostly use), and added an icon to bring up a new Enhanced Editor window. To add your own customized icon, follow these instructions:

1. Click within the window that you want to customize to make it active.
2. Right-click the toolbar at the top of the window, and select **Customize.** The Customize Tools window is displayed.
3. Click on the **Customize** tab, as shown in Figure B.1.

Figure B.1: Customize Tools Window

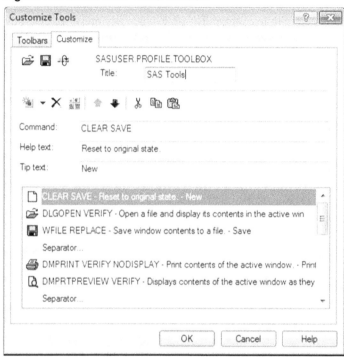

Continue by completing the following steps on the Customize Tools window:

1. Use the drop-down arrow next to the New button to select **Blank Tool**. The New button is the first icon in the tool bar above the **Command** field.
2. Add the following information for the new tool :
   Command: WHOSTEDIT
   Help text: Enhanced Editor
   Tip Text: Enhanced Editor
3. Click the Icon button to choose an appropriate icon (I use the paper and pencil icon). The Icon button is the third icon above the **Command** field.
4. Use the Up and Down arrows to move your new tool to a place that makes sense.
5. Click OK, and click Yes when you are asked whether you want to save the updates to your SASUSER.PROFILE.TOOLBOX.

You can add other customized icons that you might find useful.

Using the Program Editor

If you have to use the Program Editor (on Unix or a mainframe) or just prefer to use it, here are some tips to make it easier.

Customizing Function Keys

Function keys (the F-keys on your keyboard) in the Program Editor make many tasks much easier. Instead of having to locate a command in the menus at the top of the window, you can just press a function key.

You can define your own function keys in the Program Editor to accomplish common tasks, and you can stack commands, or execute multiple commands with a single key press. To get to the Keys windows, enter "keys" on the command line, or choose **Tools ▶ Options ▶ Keys**. A lot of the keys are set by default, but I have found that some of them are not intuitive for me, so I make some changes when I have a new installation of SAS.

Table B.1 shows the keys that are defined by default and the changes that I make on Unix. The ones that I have added or changed are marked with an asterisk (*).

Table B.1: Function Keys

Key	Definition	Explanation
F1	help	Display Help.
* F2	submit buf=xterm	Submit the selected lines of code. This is the same as using the **Run ▶ Submit Clipboard** command.
F3	submit	Submit all the code.
* F4	pgm; clear; recall	Go to the Program Editor, clear everything, and recall the last submitted code.
F5	rfind	Find again. You must first use the **Edit ▶ Find** command to find a string, and then this key finds the next occurrence of the string.
F6	rchange	Change again. You must first use the **Edit ▶ Replace** command to replace the next occurrence of the string.
F7	backward	Scroll the page backward.
F8	forward	Scroll the page forward.
* F9	toolload	Make the Toolbox window active for the current window. I find this useful when I bring up a table in the VIEWTABLE window; you either need to click somewhere in the table to make the Toolbox window active, or use this command. I use the Toolbox window to enter a WHERE clause on the command line.

Key	Definition	Explanation
F10	left	Scroll the page to the left.
F11	right	Scroll the page to the right.
* F12	clear	Clear all the text in the current window.
Shift+F1		
Shift+F2	keys	Bring up the Keys window.
Shift+F3		
* Shift+F4	pgm; recall	Go to the Program Editor and recall the last submitted code. Note that this key does not clear the contents of the Program Editor before doing the recall.
Shift+F5		
Shift+F6		
Shift+F7		
Shift+F8		
Shift+F9	pmenu	Toggle between the menus on all the windows and the command line.
Shift+F10		
Shift+F11		
Shift+F12		
Ctrl+A	next	Go to the next window. (I actually don't use this; it doesn't always bring up the window that I expect.)
Ctrl+B	prevwind	Go to the previous window. (I actually don't use this; it doesn't always bring up the window that I expect.)
* Ctrl+C	store	Store marked text on the clipboard (just as Ctrl+C works anywhere else).
Ctrl+D		
Ctrl+E	clear	Clear the text from the current window.

Key	Definition	Explanation
Ctrl+F	mark	Mark a selection. On Unix, you can't just mark a selection. You need to put your cursor at the beginning of the lines you want to select, press Ctrl+F to mark, and then click at the end of the lines you want to select.
		Note that when you want to copy a piece of code, you can click and drag it. It is then selected and stored on the clipboard so that you don't need to use this mark and store method to do a basic copy and paste.
Ctrl+G		
Ctrl+H	unmark	Unmark a marked section.
Ctrl+I		
Ctrl+J		
Ctrl+K	cut	Cut a marked selection so that you can paste it somewhere else.
Ctrl+L	log	Go to the log window.
Ctrl+M		
Ctrl+N	undo	Undo the last change.
Ctrl+O	output	Go to the output window.
Ctrl+P	pgm	Go to the Program Editor window.
* Ctrl+Q	wsave	Save the options, size, and location of the current window.
Ctrl+R	store	Store marked text on the clipboard (this is the same as Ctrl+C).
* Ctrl+S	store	Store marked text on the clipboard (this is the same as Ctrl+C and Ctrl+R).
Ctrl+T	paste	Paste whatever has been stored on the clipboard (this is the same as Ctrl+V).
Ctrl+U	home	Go to the command line, if one exists.
Ctrl+V	paste	Paste whatever has been stored on the clipboard (just as Ctrl+V works in most places).

Key	Definition	Explanation
Ctrl+W	command	Toggle the command line and menu on the current window. This is similar to the pmenu command (Shift+F9), except that the pmenu command toggles the command line on all windows, not just the current one.
* Ctrl+Y	redo	Redo the last undo.
* Ctrl+Z	undo	Undo the last change.

As you can see, there are lots of available keys that you can set. I try to set keys based on what I might already be used to (such as Ctrl+Z). There are quite a few keys that I don't bother setting simply because I don't need (and can't remember) that many shortcuts. If you want to set more, the available commands are in the documentation for the SAS Windowing Environment. Don't forget that you can execute more than one command on a single function key by just using a semi-colon between the commands.

Setting Options

There are two options that I always set for the Program Editor using **Tools ▶ Options ▶ Program Editor**:

- I turn on the **Split lines on a carriage return** option. This enables me to put my cursor anywhere on a line and press the enter key, and everything to the right of the cursor is moved to the next line. This is pretty typical behavior for a text editor, so I like to set it for the Program Editor.
- I turn on **Display line numbers**. I always want lines numbers when I'm programming, and you must have line numbers in order to use the Line Commands (explained in the "Using Line Commands" section).

After setting the options and closing the Options window, you must then use the wsave command to save the setting. The wsave command saves options as well as the size and location of the current window in your sasuser profile, so they will be applied the next time you bring up SAS. If you have created a function key to do wsave (I have Ctrl+Q set to wsave), then you can just use that to save the window. If not, then go to the toolbox or command line and enter "wsave".

Using Line Commands

If you have ever edited files on z/OS using TSO, you are familiar with line commands. If you haven't used the TSO editor or the SAS Program Editor, line commands might be a strange concept. Basically, these commands enable you to type over the line numbers of the Program Editor to manipulate the text as necessary. Some of the commands enable you to specify a number of lines, and some let you specify a block of lines.

Table B.2 is a list of the commands that I find most useful.

Table B.2: Line Commands

Insert	
i	Insert a line after the current line.
i#	Insert # of lines after the current line.
ib	Insert a line before the current line.
ib#	Insert # of lines before the current line.

Delete	
d	Delete the current line.
d#	Delete # of lines, including the current one.
dd.. dd	Delete the group of lines starting at the first dd and ending at the second dd.

Copy	
c.. a	Copy the current line after the line marked by a.
c.. b	Copy the current line before the line marked by b.
cc..cc.. a	Copy the lines from the first cc to the second cc after the line marked by a.
cc..cc.. b	Copy the lines from the first cc to the second cc before the line marked by a.

Move	
m.. a	Move the current line after the line marked by a.
m.. b	Move the current line before the line marked by b.
mm..mm.. a	Move the lines from the first cc to the second cc after the line marked by a.
mm..mm.. b	Move the lines from the first cc to the second cc before the line marked by a.
m.. o	Move the current line over the line marked by o. The current line replaces any text that is currently on the o line except for blanks. If there is a blank in the m line, the text on the o line is preserved.

Repeat

r	Duplicate the current line directly below it.
r#	Duplicate the current line # times directly below it.
rr..rr	Duplicate the set of lines marked by the first rr and the last rr.
rr#..rr	Duplicate the set of lines marked by the first rr and the last rr # times.

Indent

> or)	Indent the current line by one character.
># or)#	Indent the current line by # characters.
>>..>> or))..))	Indent the group of lines marked by the first >> or)) and the second >> or)) by one character.
>>#..>> or))#..))	Indent the group of lines marked by the first >> or)) and the second >> or)) by # characters.
< or (	Unindent the current line by one character.
<# or (#	Unindent the current line by # characters.
<<..<< or ((..((	Unindent the group of lines marked by the first << or ((and the second << or ((by one character.
<<#..<< or ((#..((	Unindent the group of lines marked by the first << or ((and the second << or ((by # characters.

Miscellaneous

cols	Display a horizontal ruler showing column numbers
uc or cu	Convert the line to uppercase.
uuc..uuc or ccu..ccu	Convert the lines that begin with the first uuc and end with the last uuc to uppercase.
lc or cl	Convert the line to lowercase.
llc..llc or ccl..ccl	Convert the lines that begin with the first llc and end with the last llc to lowercase.

If you've never used line commands, the following examples show how they work. Figure B.2 shows how to use i# command to add two lines. The results are shown in Figure B.3.

Figure B.2: Inserting a Single Line—Before

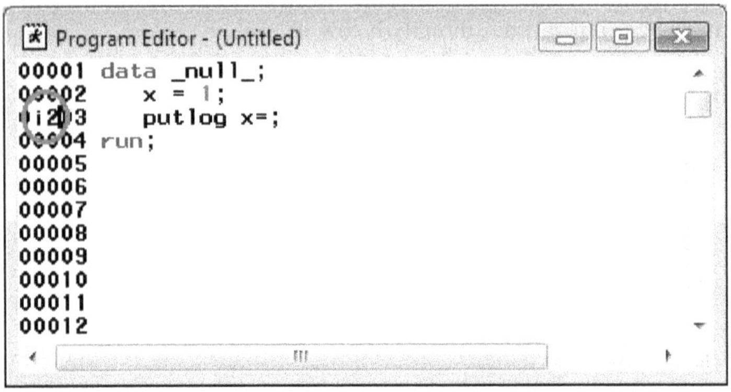

Figure B.3: Inserting a Single Line—After

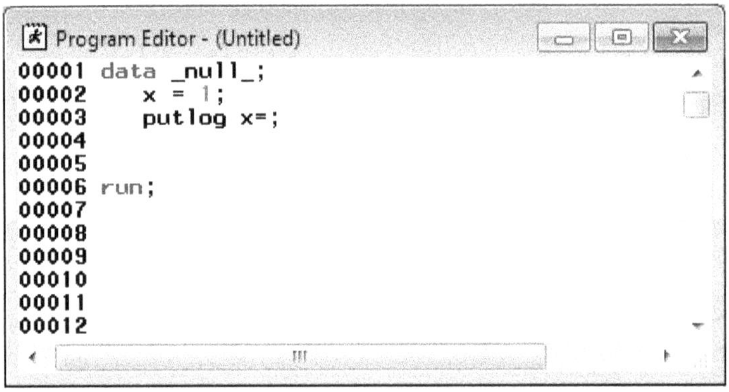

Figure B.4 shows how to use the dd..dd command to delete two lines. The results are shown in Figure B.5.

Figure B.4: Deleting Three Lines—Before

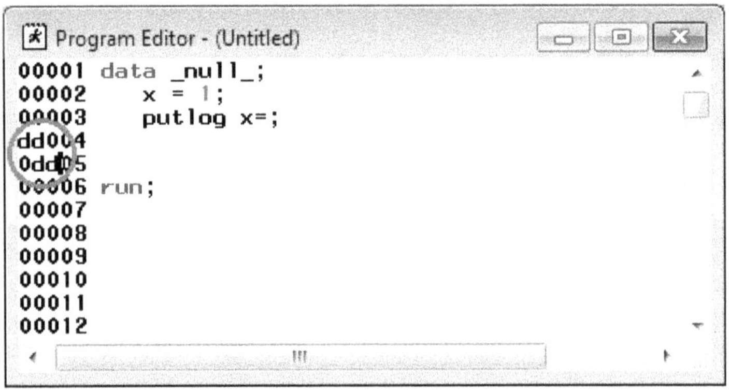

Figure B.5: Deleting Three Lines—After

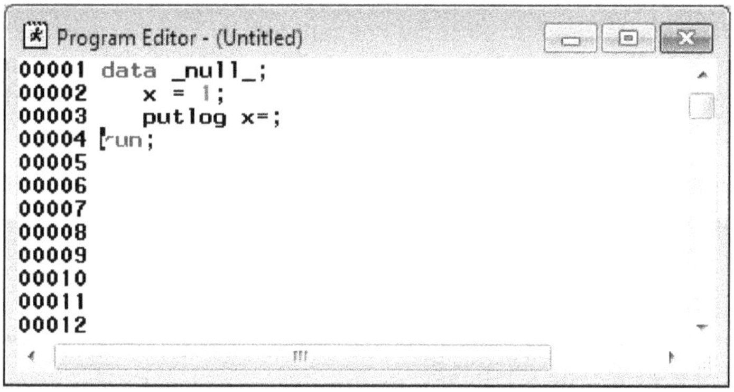

```
Program Editor - (Untitled)
00001 data _null_;
00002     x = 1;
00003     putlog x=;
00004 run;
00005
00006
00007
00008
00009
00010
00011
00012
```

Figure B.6 shows how to use the mm..mm..a command to move two lines after another line. The results are shown in Figure B.7.

Figure B.6: Moving Two Lines after Another Line—Before

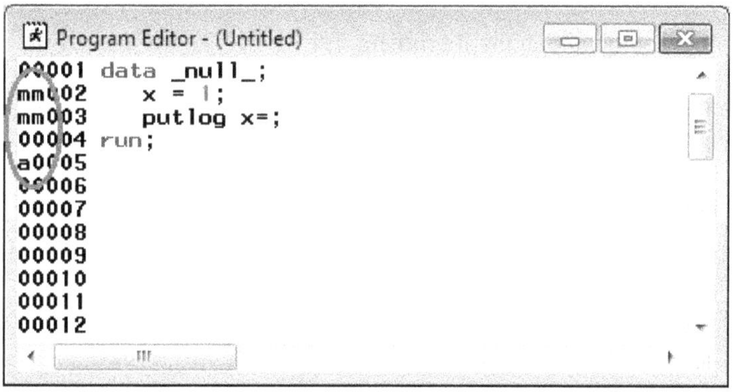

```
Program Editor - (Untitled)
00001 data _null_;
mm002     x = 1;
mm003     putlog x=;
00004 run;
a0005
00006
00007
00008
00009
00010
00011
00012
```

Figure B.7: Moving Two Lines after Another Line—After

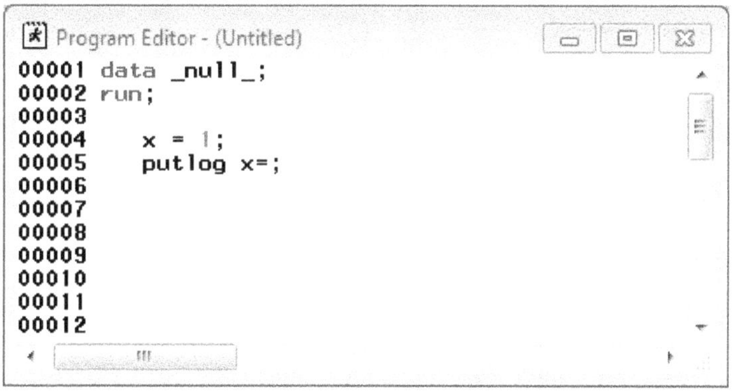

```
Program Editor - (Untitled)
00001 data _null_;
00002 run;
00003
00004     x = 1;
00005     putlog x=;
00006
00007
00008
00009
00010
00011
00012
```

Working with the Recall Stack

When you submit code, the default behavior is to clear the Program Editor. You can then use the recall key (F4) or command to bring back the code that you just submitted. You can keep using the recall key to bring back each piece of code that you have submitted during the current SAS session.

There are times when you recall too far back in the stack. When you do this, clear out your editor window, type a semicolon, and submit it. This returns you to recalling the most recently submitted code.

Handling Unbalanced Quotation Marks

Nothing is worse than submitting some code that has a missing quotation mark. It leaves you in limbo because SAS thinks that you are trying to quote everything that you submit from then on, so you can't get anything to work. When you get into this situation, try submitting this code:

```
*'; *"; */;
```

This usually clears out a mismatched quotation mark problem.

If you have also been submitting macros or might have mismatched parentheses as well, you might need to submit the following code if the first cleanup doesn't work:

```
*); */; /*'*/ /*"*/; %mend;
```

Try submitting these lines of code multiple times if they don't clear up the problem. If that still doesn't work, you probably need to shut down SAS and start again.

Appendix C: Coding Style

Introduction

I believe strongly that programmers should use a consistent style when they program. I don't care what that style is (well, I do a little bit); it just needs to be consistent. If you have a style that makes your code easy to read with plenty of white space, both you and anyone who inherits your code will have a much easier time maintaining it. If you don't use a consistent style, other programmers will waste time just trying to figure out which statements go together and when one section ends and the another starts.

I recommend that if you don't have a coding style, you should come up with one that makes you comfortable or borrow pieces from other programmers' styles. Once you know what your style is, be consistent. And remember that style is not to make your code look pretty; it is to make it easier to read and understand. Prettiness is just a nice side effect.

The following sections describe my coding style as an example.

Indenting

I indent all substatements three spaces. I have tried two spaces and four spaces (which is the default in most editors). I find two spaces to be too small (not enough differentiation), four spaces to be too much (code starts rapidly spreading across the page), and three to be just right.

It is a very controversial subject, but my choice is to use spaces instead of tabs. Many programmers believe strongly that tabs should be used to indent code, and they have many compelling arguments about it (just google "tabs for indents" to see the many arguments, pros and cons). My main argument for using spaces is that whenever I inherit a program that has tabs, the tabs have usually not been used consistently. So if you bring the code into an editor with a different size tab, suddenly the code doesn't align at all, because the programmer has some lines indented with spaces and some with tabs. If you always use spaces, your code will look good in any editor. If you are going to use tabs, then always use them.

Whatever editor I use, I set the default tab length to 3, and I set the option to use spaces instead of a tab. This means that I can still use the tab key, but it puts out three spaces instead of a tab character.

I indent any line of code that is a child of the previous line. So in SAS, I indent the code that comes after a DATA statement, a procedure statement, a %MACRO statement, an IF statement, or a DO statement, as well as others.

Here is an example (using hyphens to show the spaces):

```
%macro test;

---data _null_;
------if (weekday(today()) in (1, 7)) then
---------putlog "It's the weekend!";
------else if (weekday(today()) eq 6) then
---------putlog "TGIF!";
------else
------do;
---------numDays = 6 - weekday(today());
---------putlog numDays "days till Friday";
------end;
---run;

---proc print data = sashelp.class;
------var name age;
---run;

%mend test;
%test;
```

Aligning

Though similar to indenting, alignment has more to do lining up with other lines of code.

When a function or macro call or statement has a list of parameters, my general rule is to split the values onto separate lines if there are more than two or three values, though I am fluid on this rule. I usually decide to split the values when the line is looking too busy. So if I have a KEEP option on my DATA statement with more than three variables, I split it up so that each variable is on a new line, and I align the variables with each other like this:

```
data test (keep = numeric_1
                  character_1
                  character_2
                  numeric_2
                  date_1);
```

I also align the start and end statements for a block of code. I like to have the statement that begins the block of code line up with the statement that ends the block of code. So I make sure that a DATA statement or procedure statement is aligned with its RUN statement, a DO statement is aligned with its END statement, and a %MACRO statement is aligned with its %MEND statement. I like to be able to look down the left side of my code and easily tell where a block begins and where it ends.

```
|%macro test;
|
|   |data _null_;
|   |   if (weekday(today()) in (1, 7)) then
|   |       putlog "It's the weekend!";
|   |   else if (weekday(today()) eq 6) then
|   |       putlog "TGIF!";
|   |   else
|   |   |do;
|   |   |   numDays = 6 - weekday(today());
|   |   |   putlog numDays "days till Friday";
|   |   |end;
|   |run;
|
|   |proc print data = sashelp.class;
|   |   var name age;
|   |run;
|
|%mend test;
%test;
```

Note that my DO statement is on its own line and it lines up with its END statement. The IF or ELSE statement doesn't have an associated END statement; the END statement belongs to the DO statement. So I line up the DO and the END, not the IF/ELSEand the END. According to my own rules, I should indent the DO statement under the IF statement, but I have made an exception in this case. That's because if I have an IF statement without a DO/END block, and I then decide to add additional statements so that I need DO and END statements, I just have to add the DO and END statements. I don't need to change the indenting.

Handling Line Lengths

Years ago, we all kept our SAS code lines to no more than 80 characters, because then we could be sure that the code would work on any operating system (particularly on the mainframe systems). I don't think that rule is as important these days, especially if you are not moving your code from one operating system to another. However, for readability, I suggest keeping the length of your code lines to around 80 characters. Sometimes this isn't practical and can make it harder to read. However, a shorter line is generally easier to read, especially if a long line requires a horizontal scroll bar.

An additional thing to note with long source code lines is that the LINESIZE= option determines whether your source code wraps in the SAS log. I try not to let my code wrap, because it makes the log cleaner and easier to read.

Using Capital Letters (or Not)

I don't like to use capital letters on very many parts of my code. None of my SAS statements, functions, or variables are ever in caps.

One department that I worked in had a coding rule that any constants that are specified for a function or an option should be in caps. For example, MONTH and SAME are constants that tell the INTNX function what to do, so we would put these in caps:

```
date = intnx("MONTH", today(), 2, "SAME");
```

I'm not sure why this was the rule, but it is a habit that stuck with me.

Naming

I use lots of different styles of table, variable, and macro names, but I try to stay consistent within a project.

- My favorite style is camelCase; join all the words together and capitalize the first letter of each word except the first, like so:
  ```
  thisIsMyVariable
  ```
- My next favorite is to use all lowercase and use underscores for spaces, like so:
  ```
  this_is_my_variable
  ```
- I dislike all uppercase, either with underscores or without, simply because it is harder to type.
- For macro names, I use lowercase with underscores. This is just a habit because it is required on Unix: If the macro is going to be in an autocall macro library, the filename and the macro name must be in lowercase.
- Even though SAS is case insensitive with variable and table names, I try to stay consistent with the casing. It is just easier to always look for "myVariable" rather than sometimes seeing it as "myvariable" or "MYVARIABLE".
- Finally, choose meaningful names.

Coding with Style

Here are a few other rules I try to follow.

Using Mnemonic Operators

In expressions, I use the mnemonic version of the operators rather than =, >, or <. So I use eq, ne, gt, ge, lt, and le. This is mainly because of the equal sign. Using eq helps differentiate between an assignment and an operator.

Using Parentheses

Because I spent quite a few years going back and forth between programming in Java and programming in SAS—in many cases using Java to generate SAS code—I got in the habit of putting parentheses around my conditions in an IF statement. If I do it all the time, I don't have to stop and think whether the language requires it. I actually find that it makes the SAS code easier to read, because the condition stands out, particularly in macro %IF statements.

Using New Lines

I always put each statement on its own line. This includes putting the statement after a IF/THEN on a new line:

```
if (a eq 1) then
    b = 2;
```

This makes it much easier to add a DO/END and extra lines if necessary.

Using Spaces

I like to put spaces around operators (arithmetic, equality, and assignment):

```
var1 = var2 + var3;
```

I don't like putting spaces around parentheses; I think it causes line to be too spread out and harder to read:

```
var1 = (var2 + var3) / (sum(var4, var5) - var6);
```

I also put spaces after commas just as I would if I were writing a sentence.

Using RUN and QUIT Statements

I always put a RUN or a QUIT statement at the end of a DATA step or a procedure. These statements aren't required, because the next procedure or DATA statement signals the end of the previous procedure or DATA step. But omitting the RUN or QUIT statements is more likely to cause issues when someone adds a piece of macro code without realizing that the DATA step or procedure has not yet been terminated. It also makes the code easier to read; you can easily see the end of the step or procedure.

Here's a prime example of why this is a good practice: A friend asked me to help him figure out why a procedure wasn't setting the &SYSERR automatic variable to a nonzero value when there was an error. The problem was that he had no RUN statement after the procedure. The procedure was followed by a %PUT to put out the &SYSERR value, and then a DATA statement to start the next step. Without a RUN statement, the procedure continued to run until it reached the DATA statement. The result was that when the %PUT ran, the &SYSERR value hadn't been set yet, because the procedure hadn't finished yet.

Using Table Names in Procedures

I always use the table name in a procedure, although it is not required. The SAS program uses the last table that was created if you don't specify one. However, if a new piece of code is added that creates a table, then suddenly the procedure is no longer reading the right table.

Using Comments

If I have more than three statements in a DO/END block, I add a comment to the END statement indicating which statement the END belongs to:

```
do until(column eq "");

   ...

end; /* do until - loop through columns */
```

These end comments generally start with either "do," "do i," "if," or "else," followed by a hyphen and then a note about what the block does.

If I comment out chunks of code, then I insert a comment explaining why. However, I usually prefer to remove all the extra code before turning it over to anyone else. Source control systems are great for this. You can go back to older versions of the code if you need to, so you never have to keep old chunks of code.

Using White Space

One of the most important rules is to use white space. Add plenty of blank lines between sections and blocks of code. It is crucial to make your code readable. It helps your eye and brain quickly tell which parts of the code belong together.

Finally, all rules are made to be broken, so if I have a rule that makes the code harder to read or understand, then I abandon the rule without any hesitation!

References

Carpenter, Art. 2016. *Carpenter's Complete Guide to the SAS Macro Language*. 3rd ed. Cary, NC: SAS Institute Inc.

SAS Institute Inc. 2014. *SAS 9.4 Metadata Model: Reference.* Cary, NC: SAS Institute Inc.

SAS Institute Inc. 2003–2016. SAS XML Mapper Tool. Available at https://support.sas.com/downloads/browse.htm?fil=&cat=12.

Index

Ready to take your SAS® and JMP®skills up a notch?

Be among the first to know about new books, special events, and exclusive discounts.
support.sas.com/newbooks

Share your expertise. Write a book with SAS.
support.sas.com/publish

www.ingramcontent.com/pod-product-compliance
Lightning Source LLC
Chambersburg PA
CBHW080522060326
40690CB00022B/4997

Learn to write SAS® programs quickly and efficiently.

Programming in SAS is flexible, but it can also be overwhelming. Many novice and experienced programmers learn how to write programs that use the DATA step and macros, but they often don't realize that a simpler or better way can achieve the same results. In a user-friendly tutorial style, *Practical and Efficient SAS® Programming: The Insider's Guide* provides general SAS programming tips that use the tools available in Base SAS, including the DATA step, the SAS macro facility, and SQL.

Drawing from the author's 30 years of SAS programming experience, this book offers self-contained sections that describe each tip or trick and present numerous examples. It therefore serves as both an easy reference for a specific question, and a useful cover-to-cover read. As a bonus, the utility programs included in the appendixes will help you simplify your programs, as well as help you develop a sleek and efficient coding style.

With this book, you will learn how to do the following:

- use the DATA step, the SAS macro facility, SQL, and other Base SAS tools more efficiently
- choose the best tool for a task
- use lookup tables
- simulate recursion with macros
- read metadata with the DATA step
- create your own programming style in order to write programs that are easily maintained

Using this book, SAS programmers of all levels will discover new techniques to help them write programs quickly and efficiently.

MARTHA MESSINEO is a principal software developer in the SAS® Solutions OnDemand division at SAS. She has more than 30 years of SAS experience, most of which she has spent writing SAS programs in a variety of positions, including those in Technical Support, Professional Services, Management Information Systems, SAS® Solutions OnDemand, and Research and Development. In her Research and Development role, she worked with SAS® Data Integration Studio, IT Resource Management, and SAS® Asset Performance Analytics. Before coming to SAS, she worked for MetLife Insurance in the Capacity Planning and Performance Tuning division, where she used SAS to analyze mainframe performance data. She is a SAS Certified Base Programmer for SAS®9, a SAS Certified Advanced Programmer for SAS®9, and a Data Integration developer for SAS®9.

FREE DATA on the Web!
support.sas.com/authors

SAS® Press

ISBN 978-1-63526-023-6

sas.com/books

9 781635 260236